M000099404

The Best Place for Garbage

The Essential Guide to Recycling with Composting Worms

By Sandra Wiese

WiR Farm, Bennett, CO

Copyright © 2011 by Sandra Wiese

All Rights Reserved.

No part of this book may be reproduced or transmitted in any form or by any means, electronic or mechanical, including photocopying, recording, emailing or by any information and retrieval system, without permission in writing from the author.

Published by:

WiR Press

P.O. Box 415

Bennett, CO 80102

ISBN-13: 978-0-615-43158-1

ISBN-10: 06-15431-585

The author and publisher do not assume and hereby disclaim any liability to any party for any loss or damage caused by errors or omissions in this book or by the readers' own negligence, errors or omissions or other cause in applying the information contained within this book. The author further disclaims any responsibility for your friends making bad worm jokes and ridiculing you for saving garbage. She will join you, however, in peals of giddy laughter when later these same friends ask you to teach them how to raise worms themselves.

Cover art by Jenna Kolb

jenkolbdesigns.com

This book is dedicated to the pioneers who bravely march on
despite all odds and ridicule.
The same ones who turn the weird and impossible into everyday habits
for the rest of us

.

CONTENTS

ACKNOWLEDGMENTS

Every book acknowledgement I've ever read has a huge list of folks whose "help was invaluable", "it takes a village", "I could never have done it without" blah blah blah. No one ever says, "Hey! I did this all on my own! So everyone else can eat my shorts!" Well, sorry, you won't find anything different here because in this case it actually did take a small tribe. First thanks to my friends for all the jokes and laughs about raising worms, even though most of you now have seen the light and raise them yourself. Next to all my students who asked for this time and time again…Here you go! To all those who read it first, friends and strangers alike, who declared this book "worthy". Huge, huge, ginormous thanks to Angela B., Steve C., Addy E. and Melissa T. for putting up with my endless whimpering and questions. Endless ginning over the cutest worm ever goes to Jenna Kolb, designer extraordinaire. Most of all, thanks to the Christ Jesus, without whom I am nothing. Props.

Why Raise Worms?

Welcome to the Best Place for Garbage! After teaching worm composting to people of every sort of background, I've found that they all have at least one thing in common: everyone who wants to raise worms for the first time has questions that beget questions that turn into even more questions, most of which they didn't even know they had to begin with. It is my hope that this book can not only provide answers to all of those questions, but perhaps answer some questions *before* you even know you have them. I've tried not to leave anything out, including humor. After all, this book is about garbage and worms and poop – subjects more ripe for joking than most.

When I first started out wanting to raise worms, I found a ton of information on raising composting worms on the internet and in the few other books there are available on the subject. Great, right? It turned out to not be so great after all. So very much of the information I read was contradictory. Some claims seemed ridiculous. A whole lot of it didn't stand up to the light of logic. But I muddled my way through and by actually *doing* I found out that many of the contradictory things were actually completely untrue and many of the things that seemed illogical on the surface turned out to be the same way deep down. There are so many options when raising worms at home and so many different perspectives, I decided to put them all in one place, but in this case also including the realities found from actually raising worms to eat garbage, not from rumors I heard from someone else.

Who the heck am I? The simple answer is that I am just a modest worm farmer with a crazy dream that some day "I have worms" will not be something that makes the listener think: Well I hope your doctor can give you something for them!

Reader Warning: I started out with just two meager pounds of worms. Just two. As of this writing, I have about 300 pounds. Yeah. Three hundred. I'd have more, but I don't just raise them, I sell them. In the world of commercial worm farms, I'm *small*. Very, very small. So what is the warning? *It could happen to you.* Nope, I'm neither insane nor am I kidding. Remember: the crazy cat lady from your childhood also likely started her journey with just *one* cat.

I also teach classes on how to raise worms. I get all kinds of people in the class and there is one thing they all have in common: most of their friends already think they are weird. I mean really, they are taking a *class* to learn how to grow worms. One thing I can always tell is who I'm going to get the most questions from *after* the class.

If you are not the nurturing type, if all things green do not already thrive under your touch, if baby animals do not inexplicably recover with your careful ministrations, if people far and wide are not miraculously cured by your homemade soup, then you will have a LOT of questions. Worms are living creatures and require a certain level of care. Some folks are more cut out for this and will find the whole process rather intuitive. Some are, well, not. Don't worry, this book is for both of you. I try to cover *everything*. If you are the "intuitive" type, a lot of this will seem obvious and natural. If the process doesn't seem natural then please don't loan this book to anyone because you will need it for reference. If you want to share your newly acquired worm knowledge, then by all means buy your friends this book as a gift instead, just don't let your own copy out of your sight!

Worms are oddly and inexplicably addictive. Prepare yourself, even if you only have just a pound or two, to hear gales of laughter and bad, bad jokes from your friends. You may choose at the start to never ever never tell anyone that you have worms. That is totally okay, no harm, no foul. But the unfortunate reality for you is that likely you *will not be able to stop yourself* from spreading the worm...I mean word. See? Bad worm jokes abound.

If you have gotten to the point in your recycling endeavors (and that is what leads so many to vermicomposting, the technical name for what you are about to do) that you are seriously considering allowing *worms to live in our home* then you are gone. Overboard. A lost cause. Your friends and family have already heard all about reducing, reusing and recycling from you, so likely you won't even be *able* to keep your vermi-habits quiet even if you think you want to. Good luck trying.

Frankly, I hope you fail miserably about keeping quiet. In fact, I'm praying that you do not. Our only hope for acceptance in this unaccepting worm-less world is to spread the word so far and wide that it becomes abnormal to *not* have worms. Imagine this conversation of the (near?) future:

Wormer: Um, pardon me, but did you seriously just throw those coffee grounds in the *trash???*

Future Wormer (Everyone <everyone!> is a future wormer. There are no non-wormers): Duh, of course I did. It's *trash.*

Wormer: Oh, no, my fine friend. It is not only NOT trash that you just threw away, it is *black gold* that you just threw away. An environmentally sustainable, soil enhancing, plant loving, humanity saving, filter wrapped *goldmine* you so casually put in the refuse bin.

Future Wormer: Oh my gosh! I had no idea. Tell me more!

(At this point you use your new found knowledge of raising worms to eat your garbage and the tremendous and virtually limitless benefits they provide to kindly, humorously and informatively educate this Future Wormer in the way things really ought to be. Your inspiring delivery of said knowledge impels said Future Wormer to immediately decry for all time the

3

useless waste of recyclable and reusable goods and begin raising worms for their ownself.)

Sigh. A girl can dream, right?

But if you never start...Onward with the gaining of knowledge! I do hope you find this humble tome worthy of your time and more than a laugh or two. I wish you every success in your modest and often not-so-modest efforts to better this planet we share and your lovely little spot on it.

Reader caveats (there are only two!):

1) You will find in your research that some of what I say fully or slightly contradicts something you read somewhere else. If this is the first thing you have read about worms, further research will find someone disagreeing with me (how dare they?!). This book is about raising worms to harvest their poop, it is not about brain surgery. There are many ways to successfully accomplish this goal. There are also many ways to totally screw this up. What you have in your hands is everything that I know about what *works* and everything I know about what definitely doesn't work; follow it and you will have success in your quest for worm poop. You find a way that you think is better? Excellent. I can't wait to read *your* book!

2) I have been told by a few, thankfully few, people that my liberal use of the term "worm poop" to describe the end products of this process is very offensive and childish. I even got "appalling" once. (No, really. I could not make that up.) It is true that there are more delicate ways of saying it: vermicompost or worm castings come to mind. If the term "worm poop" is offensive to you, please *stop reading immediately if not sooner!* The much sought after end product of this process IS worm poop. All other terms are really just putting lipstick on a pig. If you want to hear about this from someone who won't tell you the bare truth then please by all means do so. I don't care where you learn about this process, as long as you DO. Everybody poops and some of

that poop is quite valuable. So let's get moving so you can find out how to get some valuable poop for your ownself.

Section One: Getting Started

-worm real estate
(bins and such)

-furnishings
(proper bedding)

-put a lid on it
(necessary cover)

-cool breezes and worm rain
(proper air and moisture)

-worm chow
(what they eat)

-to infinity and beyond
(bin maintenance)

Worm Real Estate:

Types of Worm Homes and Bins

Like all of God's creatures, worms need a place to call home. Generally, the ground serves this purpose quite well. Unfortunately, if you want a supply of worm poop for your own personal use (or to give as a gift, but more on that later), then you really only have two choices: 1) you can get them out of the ground and into a controlled environment or 2) you can spend your days trolling the yard for worm turds with a magnifying glass and a pair of tweezers. Presuming you prefer option one, your "controlled environment" has a few basic requirements, but the short story is that anything that is big enough and will hold water will also hold worms.

Generally, regardless of the type of container used, your worm home is referred to as a "bin". I have worm bins of all shapes and sizes. I use plastic storage containers, cut up barrels, old recycling totes, an inside-out tractor tire, old two gallon water jugs and even several old dresser sets that were given a new life after they became too ugly and old for other storage. You may also find several free or low cost plans to build a worm bin on the internet (including a worm bin made out of a pair of old jeans). I know for sure you can find several different commercially made bins available for sale.

My personal philosophy and one of the greatest things about worm composting is that this whole process is more about recycling things that should never be put in a landfill (i.e.: your kitchen scraps and old papers and such) than it is about the end product. Don't get me wrong, I'm an avid gardener so the worm poop really is worth its weight in gold to me. In a nutshell, I am very much against buying new things for your worms and very much for reusing what you have handy and not spending money or resources on your worms. It isn't just that I'm cheap (though I AM), it just doesn't make any sense to spend a lot of money to recycle something when what you have on hand will almost always work just as well.

Vermicomposting is very much what I like to call an *uber*-recycling endeavor. Composting is one of the only types of recycling that you can do and get an actual personal benefit from. You can take all the plastic water bottles you want to the recycle bin, but try to turn them all into something useful on your own and you are pretty much out of luck. So to be participating in this hugely beneficial process and to be keeping all these organic recyclables out of the landfill just doesn't jibe with buying some fancy worm bin or, of all things, commercially prepared worm food (no, I'm not making that up, it exists). That said, if you already bought some fancy commercially made worm bin, don't fret about it too much. It happens.

Regardless of the type of container you choose, there are some basic rules to follow. First, unless it is a commercially made bin and they already did it for you, I do not recommend drilling any holes in your container, especially not in the bottom. I'll cover "why not" soon...Second, though it seems obvious, make sure your container is moisture worthy. Some things may not seem moisture worthy on the surface (like a clothes dresser or a file cabinet), but they can be made worthy with very little effort. Effort we'll talk about as it comes up. Third, you will generally want some kind of cover. An actual lid is great, but you can rig up a cover with just about anything. Yep, we are also going to cover that (no pun intended) in a bit. Lastly, make sure it is the right size. Generally, a pound or two of worms is a good starting place for home vermicomposting. You will want to go by the general rule of thumb of one cubic foot of volume (space) per pound of worms when deciding on a bin size. That's cubic foot, not square foot. That said, you don't necessarily want a square bin that measures one cubic foot. Oh, stop, I am not being confusing, just keep reading.

What the "cubic foot" rule means is that it should be able to hold approximately that volume of bedding. We will cover bedding specifics in the next section, but you will not want your bedding depth to be too much less than six inches or much more than eight inches deep. Deeper bedding presents its own problems, most of them super smelly. You will make up the volume requirements (to equal your total of one cubic foot of space) by expanding the length and width of the bin. No, you do not need an engineering degree to figure this out, I swear. It is a rule of thumb, not one of the Ten Commandments so don't over-think it or try to be too precise about it. Just make the height of the bedding times the width times the length equal about one cubic foot of space.

To make it as easy as possible, try to find, make or improvise a bin that is about the size of four regular-sized shoe boxes put together, only a couple inches deeper. Much smaller or too shallow and maintenance becomes a pain. Much larger and the whole process slows down because larger bins will alter the worm to space ratio in an unfavorable way. Plus, if it is too big you will have a very hard time moving it around, and you likely will need to move it around at some point. A cubic foot per pound of worms will allow room for your worm population to increase and still be small enough to manage easily.

When choosing the type of bin you will use, keep in mind where it is you want to put it. Remember though, that sometimes where you think you want to keep the bin and where it ends up actually being are not always the same thing, so keep an open mind about the placement, especially if you have a spouse or housemates that are none too keen on this whole idea to begin with.

Let's go over some of the pros and cons of different things you can use as a bin:

Plastic Storage Bins, Barrels and Buckets

By far the easiest and likely also the most common type of worm bin is a plastic container. Most everyone has some type hanging out around the house that they can use. I'm thinking of the kind that rhymes with Blubber-maid or Girl-ite. If you don't think you have one, this is the perfect time to sort through all those holiday decorations or old sweaters and donate the items you no longer use to your local thrift store. Ta da! Now you have an empty container for your worm bin. You absolutely can use one of those mondo-sized, two foot tall tubs, but keep in mind you will only be using the bottom third for your worms. You can also absolutely use the "clear" kind. And always remember that the bin and contents will weigh about ten times more when it is finished and filled with worms and worm poop, so don't use anything that is too large if you are not strong enough to move it. More importantly, don't plan to place it anywhere above the height of your chest or you are just asking to be bonked hard on the head with a bin full of poo (not, of course, that I had to learn this the hard way). So really, it is better to find a bin in a manageable size.

Many websites and such will advise you to drill multiple holes into the sides and bottom and lid of your bin to allow for air flow and drainage. I say NO! First of all, this is too high maintenance for me. I'll say it now and again and again: I'm essentially lazy. Second, if you decide that raising worms is not for you or you change your bin type down the line, you will be stuck with one of those nice and sturdy plastic containers that is now filled with holes and useful for very little else because of these same holes you spent all that time putting in your bin. You may then be tempted to toss it and pretty much negate any of the world-saving good you might have done with your worm composting efforts.

One of the main reasons you will be advised by others to drill holes in your bin is for drainage. What this means is that liquid, known as leachate in the vernacular, will be able to drain out and you can collect it and water your plants with it. That is totally true. What they don't tell you is that you may have to do this *every single day*. If you don't, the liquid tends to, um, well, stank up your world with an unholy odor. Do you really want something *else* to add to your "do every day" list? The bigger problem is that, these holes are a pain in the neck with southern aspirations if you end up with flying insect problems. (I'll tell you how to deal with those when we talk about the potential roommates your worms may have in your bin.)

If you do have a lid that goes with your plastic container, great. If not, that is easily dealt with. How you deal with getting air into your bin and keeping moisture levels reasonable depends in part on where you are going to keep your bin and if you have house cats or litter trained ferrets or pet rats or any other kind of otherwise friendly household critter that will want to use your bin as a fancy new litter box or try to eat up the goodies you are feeding your worms. (As my Labrador retriever likes to say: Is that rotten food? No, it's not bad, lemme see it. I just want to roll in it! Just a little. And then maybe eat it. Just a little. Then maybe throw it up all over the couch. Just a little.) Covers for any type of bin will be discussed in its own section, but this is a great time to talk about exactly where you are going to put your bin.

In my experience, most folks like to keep the container right in the kitchen. Under the sink is a great place to put it. If you want to grow worms it's even pretty likely that by now you have replaced the myriad of chemical cleaners most folks put under the sink with just a few eco-friendly cleansers. If not, now is a great time to decide how you want to contribute to making the world a better place. Worms, of course, are a great start, but maybe putting them under the sink if you still have a mess of chemical cleaners is not the best idea.

The reason people like to put the bins in the kitchen is so that they can put the food scraps right in as they are generated. This does make life simpler, and I am all for that. We will talk about food more in a little bit. Putting your scraps right in the bin as you come across them is totally fine, but I do have a bit of a better way to do it for people with indoor bins.

Where you don't want to keep your bin is important, too. Don't put it where the dog or cat can get at it. Or the pig, if you have one of those indoor-type pot-bellied pigs. I don't recommend that you put it on top of the fridge, since it will eventually get heavy and few folks are taller than their fridge, so putting it there violates the "not above chest high" rule. I definitely don't recommend that you put it in the basement bathroom or junk room where you hardly ever go because when life gets busy, you will forget that they need your attention. Some folks like to keep them in their main bathroom (presuming you have more than one). They are there every day and can check on maintenance, etc. Under the bed is fine, too, as long as the pets can't get to it. In the closet...Essentially you can put the bin anywhere in your home where you will 1) remember it and 2) they will be in the kind of temperature range where you yourself would be comfortable.

Worms like it in the 70's, just like me. But they will live and eat and reproduce if the temps are in the upper 40's to the mid- to upper-80's. Much colder and things slow down too much. You don't want a plastic bin to be outside where it gets below freezing since your worms will freeze right along with everything else. They are, after all, made up of mostly water. Likewise you don't want the bin where it gets too hot. Because of the level of moisture in the bin, however hot it is outside it will be even hotter inside because of the natural humidity in the bin. You know the old boor's observation: it's not the heat, it's the humidity? Well, if the old boor were telling that to the worms, they'd be nodding their little pointy heads cause they would totally agree.

Which is why I don't recommend that you put them in your garage? Most garages are too hot in the summer and too cold in the winter. It's okay to put them out there in the spring and fall, but the rest of the time find a better spot without the temperature extremes. I have a friend who is one of the garage worm bin exceptions: she lives in a home that is surrounded on two sides by other people's homes. As is her garage. Her garage never gets super cold and never gets super hot. Her worms are always at a pretty even temp, give or take ten degrees or so. Essentially, her neighbor's homes heat her worms. Sweet! Except when the temperature is below freezing for more than two nights in a row. Then it is time to bring the worms inside, because apparently even a garage surrounded by other people's homes is still susceptible to getting frozen, an unfortunate lesson we both learned the hard way.

Also, if you are going to be putting your worms on a concrete floor and you do live in a place where the temperatures approach or dip below freezing, make sure you put your worms on a couple of layers of old carpet or carpet pad or a folded up old sleeping bag or the like to keep it from freezing from the bottom up. If you are going to have cold temperatures for more than a day or two, wrap the outsides of the bin in an old blanket or two, also. Just make sure you don't block off your worms air supply. You can even keep that arrangement for the whole of winter in most semi-chilly-but-not-frostbite-territory climates.

You also never want to put your plastic bin someplace where the sun will be shining on it, especially if it will be shining on it for several hours a day. If you do, you will be wishing worms were tastier as they will be thoroughly cooked. Putting a worm bin in the sun will have the same effect on the inside of the bin as it has on the inside of your car when you park it in the sun with the windows up. Wherever you decide to put a bin like this, remember it is just plastic. It doesn't have insulation and the worms are at your mercy.

Other types of plastic containers can also be used. The top foot and bottom foot cut off of any plastic food grade 35-or-so-gallon drum are great (make sure the top still has the stopper in it so you don't have a big hole when the top becomes the bottom). You can also use these same barrels cut in half the long way if you have enough worms (remember you won't be able to move something like this when it is full so place it carefully). You can certainly use wooden barrels the exact same way.

Many people try to use five gallon buckets and/or kitty litter buckets. This works. And it doesn't. Any container like this that is deeper than it is wide but so small that you have to use all or most of the available space has lots of potential for going sour (sour = smelly in the worst way) and as such I do NOT recommend you use buckets. You can, in a pinch, use either of these if you have a lid. Just caulk or otherwise seal the lid to make it waterproof, lay the whole shebang on its side, then grab your handy hacksaw or reciprocating saw and lop off the top third, lid and all. The top third will now be your lid and the bottom two-thirds will now be your worm bin.

Plastic or other garbage cans will pose the same sourness problems on a much, much larger scale. Due to their size and shape, these are generally more difficult to use the reciprocating saw trick with and most of the modern ones lose their structural integrity when you cut them up and fill them with wet, heavy stuff. For these reasons I do not recommend garbage cans in any incarnation. At least until you really know what you are doing, and then you will find yourself imagining all types of improvised bins. It's kind of cute when this happens, and maybe a little obsessive. Sorry, there is not much you can do about it. By this point accept the fact that you are worm-obsessed.

No matter what type of bin you finally decide on, if you are going to keep it in your house in a common area, like the kitchen, make sure it is something you can easily clean. It will sometimes get a tiny bit dirty on the edges or handles when you are doing your maintenance. You definitely want something you can easily wipe off with a damp sponge. Especially if someone else in your house is looking for a reason to get rid of the worms.

Now, before you read the rest of this section, please (PLEASE) start your first worm bin in one of the plastic containers I mentioned first. They are by far the easiest to maintain and move about if needed and as such will give you a very good idea of how this whole process works. After three or four months, you will be enough of an expert and *then* you will have a better idea how to make some of these other worm homes work well for you and your situation. I want this to work for you. I want you to be happy as a clam with the whole concept and process and tell all your family and friends and even strangers on the bus all about it. One day, everyone will have a worm bin just like one day everyone will recycle all of their other trash...Hey! Everyone needs to have goals, right? But I need your help and starting with a simple plastic bin is one way I know for sure to get you closer to my goal.

Wooden Bins, Dressers and File Cabinets

Plans and drawings for how to build a worm bin from wood abound on the internet. Some people will even charge you for them. If you need actual *plans* on how to build a *box*, chances are you deserve to get charged for it. It's a BOX. If you do build a wooden bin and want to use glue to hold it together, be very sure your glue is both water resistant and **in**organic. Worms LOVE to eat adhesive and will work at your glue day and night until the whole thing falls apart. I do think wooden bins can be attractive, but personally I do not make them because in order to make them sturdy enough to last, you have to use some decently thick wood and sometimes the bins can get too heavy even when empty. I do use dresser drawers, which we'll get to, but I know eventually they will fall apart. I don't mind, because with the dressers I use, I don't have to take the time to actually make them (or spend money on them). When deciding what pieces of scrap wood to use to make a wooden bin, make sure you don't use newly pressure treated wood. The chemicals used to pressure treat the wood are not good for your worms and not great for the plants the vermicompost will go on, either. You can use wood that has been painted or stained as the chemicals are generally less volatile when dry and will be mitigated by the positive effects of the worm poop (totally more on that later for sure).

If you do want to make a wooden bin, try to use a piece of laminated wood or leftover countertop for the inside of the bottom if you can. The sides could be laminated also, but this is not as crucial as the bottom, which will be essentially sitting in moisture for the life of the bin; a bin life that will be very short if you don't protect it. You can also use plastic sheeting (at least 4 mil), which is what I use to prepare the dressers and file cabinets I use.

Solid wood or sturdy pressboard dressers can be had for five whole bucks or for **free** every day on Craig's List (www.craigslist.org) and most days on Freecycle (a Yahoo group) or whatever other "keep it out of the landfill" site you like to use. Or just ask your friends if they have any. You would be surprised what junk hoarders your friends are. Unless a dresser is super ugly, you may not want to tell the person you are getting the dresser from what your nefarious plans are for it. For some reason, people can get a little weirded out and possessive if you tell them you are going to put worms in the precious dresser they heretofore did not even want in the first place.

This is one of those situations, though, where beggars do need to be slightly choosy. You don't want any dresser that will not withstand the weight of the finished vermicompost when full. If you touch it with your finger and it wobbles like mad even when empty, you will not want to fill it with worms and bedding. If it is very difficult to open and close the drawers even when they are empty and/or if the drawers do not seat well in the runners, give it a pass. Most importantly, if the drawers themselves are not jointed securely or especially if any of the sides or bottom is made of **any** kind of cardboard (thicker pressboard is totally okay), say goodbye and move along. You also don't want a dresser with all tiny underwear drawers (is that redundant? Underwear = drawers, get it?) or all very big drawers, just normal sized drawers (again: about four shoeboxes big). All of these criteria will matter hugely to you as your bin progresses since it is both a "moist" process and since the finished product will make your drawers very heavy. If it sounds like I'm being too restrictive and you will never find a free or almost-free dresser that meets this criteria, think again. There are lots. Thousands. You will have plenty to choose from. Oh, and make sure you can get it in your car if you don't have a truck or they won't deliver it. If you don't have a truck, then you already know how many people are happy to deliver just about anything to you for ten bucks or a homemade pie.

Filing cabinets also rock as worm homes. The same sturdiness requirements as for dressers also apply here, but with one other "look out for" on a filing cabinet: make sure the drawer sides are at least six inches tall all the way around. Quite a few cabinets have virtually non-existent interior side walls with only the front and back being the height of the actual drawer slot. Find one with drawer sides at least six inches high, eight is better.

Dressers, filing cabinets and homemade wooden boxes usually need a little bit of prep before they are suitable for a worm home. The bottoms, and to some extent the sides, of anything made of unprotected wood will fail over time as it will be holding moist bedding. Moisture + wood = soggy bottoms that will drop out all over your toes. Many but not all filing cabinets have holes in the bottom and sides that will need to be covered lest your worms drop out and can't get back in.

This is easily fixed with a scrap piece of Plexiglas™ cut to size or a sturdy piece of flexible plastic sheeting. For the sheeting, you will want something at least 4 ml thick. A plastic garbage bag will not do. Plastic picnic-type table cloths usually work as long as they are a decent thickness and are not a "dollar store" find. If you have a fabric store in your vicinity, you can get the thick plastic for generally less than $4.00 a yard and the roll is usually about eighty inches long so it's a good buy (better if your fabric store has weekly 40% off coupons like mine does). Cut the plastic or Plexiglas to fit into the bottom of the drawer. You can also put plastic on the sides, but this is not entirely necessary, the bottom is what will give you the most problems if it is not protected.

After you cut the plastic or Plexiglas to size, you will need to secure it in place. If you don't, the worms will get under it and will not be able to get back out and they will die. When they die, they *smell. Badly.* Secure the plastic to the bottom with duct tape. I mentioned before I am cheap, but in this area you cannot afford to be. Do not use cheap duct tape.

Worms consider the adhesive on cheap duct tape to be worm chow and will promptly eat it away. I'm generally not a brand name slave, but for the duct tape I use in my dressers I am. The only brand I have found to date that works in the long term for this use is Gorilla™ brand duct tape. This tape is widely available and while I have not tried every single other brand of duct tape, this one I know for sure the worms will not eat. It is worth the extra couple of bucks. A properly prepared dresser or file cabinet that is structurally sound will last through at least a couple of years of constant use in this manner.

Commercially Made Worm Bins

Yes, yes, I know what I said about fancy commercially made bins. I do totally stand by that. I also know that some people either already bought one or have their hearts set on getting one, so you might as well know the pros and cons.

Commercially made plastic worm bins are generally all of the same type of design. (Outside of the United States, commercially made flow through type "bins" made out of conically shaped waterproof fabric are also common. I am not very familiar with this type of worm home and have never used one, though they seem like the concept would work decently enough.) Although commercial plastic worm bins are sold under many different names, the concept for each is generally the same. Three or five square or round trays are provided, depending on how many you want to pay for. Each tray has holes in the bottom, much like a thick screen, and the trays are built to nest inside each other. The bottom of the set up has a solid tray just a couple of inches deep. This tray is generally sloped or angled at one end to meet the bottom of the lowest screened tray and has a spigot on the opposite, lowest side. This is all topped off with a fitted lid. This whole set up will put you back eighty to a hundred bucks, depending on how many trays you get. Oh, plus shipping.

The idea is that you fill the lowest holey tray with your bedding, food and worms. The worms chomp and chomp all that down into fabulous worm poo. At some point when this tray is almost done, you fit another tray on top of the almost finished tray and fill the new tray with bedding and food. The whole premise is based on the worms finishing the available food in the bottom tray and then migrating to greener pastures in the tray above. You are supposed to repeat this process until each tray is used.

Depending on which bin system you purchase, the directions will tell you the first (lowest) bin, will be completely finished and ready to use anywhere between 2 and 6 months. The spigot is used to either drain off excess water and/or to allow you to purposefully add extra water to generate compost tea, also known as leachate ('Compost tea' is generally made on purpose while 'leachate' is usually just considered a by-product of excess moisture. Essentially they are the same thing. We will discuss compost tea and its uses in detail when we talk about using the finished product at the end of the book).

Now, these types of bin systems all sound utterly fantastic, don't they? The commercially made bins do, in fact, have the supreme advantage of being generally much more aesthetically pleasing than pretty much anything you throw together in six minutes your ownself. What they don't have is a system that works *quite* like they proclaim it does in the handy instructions that come with every bin. Not to discredit those always-too-brief instructions, mind you. The instructions definitely have enough to get you started and troubleshoot a lot of the more common problems. But if they made the instructions too detailed, it would be a whole book and not just "instructions" (and frankly, my book is better). Seriously though, if they told you what I'm about to tell you about their product, you might not be so eager to plop down your hundred bucks quite so fast. I have people in my life whom I like and respect who also sell these bins. If they actually use the bins themselves, they won't ask me to delete this next little bit...

There is actually only one real reason that these types of bins don't work quite as well as advertised: worms aren't known for their overwhelming intellectual capacity. I mean, let's face a basic fact here: we are raising them to live in their own poop. Do I really need to elaborate on that?

One seldom discussed fact of worm behavior is that they totally dig rolling around on something moist and plastic. Nope, I don't have the faintest idea why. I imagine it simply feels good for them. But they aren't always smart enough to get away from something that feels good but could kill them.

The sloped bottoms of these bin systems are there for a really good reason. For one thing, this allows any excess moisture to collect on the spigot side when it comes out of the bottom of the holey trays. But the holes in the bottom of the trays also allow the worms to scootch down to this very bottom, solid tray. And boy do they! They love it! The other reason the bottom is sloped is supposedly to allow the worms to crawl back up into the bottom bin when they are done rolling around on the yummy moist plastic bottom.

And therein lies the problem. Worms don't have a superior sense of direction. No built-in GPS to tell them how to get back to where they need to be. They know to leave conditions that are bad for them, but they don't know exactly *where* to go to find better ones. They hunt and hunt and hope to find a place before they die. Sometimes they do. Sometimes...not so much. (That's where the dried worms – worm sticks – on the sidewalk after a big rain come from: worms looking for another place to be and not finding a good place before their expiration date.) Sometimes the worms find their way up the slope and back into the lower tray. Sometimes they find their way into the pool of leachate by the spigot. They can live in water for quite awhile, but not forever. They can live without food for awhile too if they have air and moisture. But also, not forever. So once a week or so, you have to make sure you take out the lower tray and scoop those silly-billies up and put them back in their tray where they belong.

Oh, "big deal" you say. "That's easy, I can do it when I feed them every week and everything will be just fine and dandy." Well, yes and no. If you forget, they'll die down there. I may have mentioned that dead worms stink to high heaven. They also tend to clog up the spigot when you try to drain the leachate. You have to try to clean them out with a toothpick or cotton swab. Oh joy.

Let's pretend, though, that you always remember. The tray with the bedding and food in it weighs about a whole two pounds when you first make it. Add a pound or two of worms and you have max four pounds. Then, the worms work their fancy magic and turn the food (which you are continually adding) and bedding (also being added) into fabulous worm poop. "Wonderful!" You say. "Exactly as advertised!" You exclaim. Except that they don't advertise that you have to do this maintenance step *at all*. And since they don't advertise that the worms love it in the bottom solid tray, they also don't add that the underside of each tray is covered in a layer of worm poop, or that a tray of finished worm poop weighs more than the initial three pounds. A lot more.

What this means to you is that when you take off the lower tray to remove the worms, you get to find a place to put the lower tray where it won't leave a tray-shaped poop outline on your floor. It also means that by the time you get to the third or fifth tray, those trays all weigh a TON. Okay, not an actual ton, but a lot more than is convenient to keep you happy with your worms and to make you want to remove the trays every week in order to herd the worms back to where they would have stayed in the first place if it weren't for the holes in the bottom of the trays. Fun, fun, fun. Make sure you call all of your friends over for that party.

The other problem I have with this type of system is that they advertise that the worms will migrate up to each successive tray if you keep feeding the upper tray. This happens. And...it doesn't. First of all, it does not take "a couple of weeks" as most of the advertisements will tell you. The majority will migrate in about a month or so. However, a large number of the very young worms will not. They just don't. Maybe they aren't strong enough. Maybe they are even dumber than grown-up worms. Hey, that's how it is with us humans, right? So they hang out in the "finished" tray until they are big enough to migrate up a tray. Or until they find that sweet, sweet moist and foodless bottom tray.

If you are doing it right, during this whole process your worms will be having loads of worm sex and making tons of cocoons (worm eggs). These cocoons will also be in the "finished" tray. Where they will hatch, making more baby worms. Believe me, if young worms are dumber than adult worms, baby worms must be about as stupid as stupid comes. Plus, they're babies, so they don't have the strength yet to push their way through all of that heavy worm poop to get to the tray you want them to be in. In any case, this "migration" doesn't totally happen the way they say it will. Ever.

There are a ton of "plans" on the internet that show you how to replicate these types of systems for much less money using regular plastic tubs. Same issues still apply with the homemade types. All that said, these systems do eventually give you a lot of poop and they look pretty on the outside. So if you bought one, by all means use it. But if you decide at some point that it isn't all it's cracked up to be, please sell it on Craig's List or some such. It's bad enough that you have a large piece of non-recyclable plastic that has no other viable purpose in life, please don't let it languish in the back of your shed.

Outside Worm Bins, Windrows and Trenches

More than one person has told me all about the box Grandpa had outside with worms in it. He'd put kitchen scraps and old leaves inside and it was always the last stop before leaving for the local fishing hole. Grandpa's worm box only comes up when I talk about what does and does not work for outside containers. Only one person I've talked to has ever had any real recollection of the box other than that it was made of wood and it was located "outside". They say he had the same box for years and years and years and it was always outside. Well, I absolutely believe that Grandpa had a worm box. I think it would be a safe bet to say it was made of wood. But the rest, well, that depends. Mostly it depends on the climate that Grandpa lived in and where that bin was kept in relation to the summer sunshine and the winter cold.

You can absolutely make a wooden box and put it outside and have your worms thrive in it. If! If it is protected from weather extremes of very cold and very hot. Very hot is worse for worms than it is for you. Worms have to have moisture to survive. Moisture + heat = humidity. If you think it is hot enough to fry an egg on the sidewalk, your worms will think it is Miami in the middle of a record heat wave. Even if it is nice weather, if a wooden bin is in the sun for much of the day it is much hotter for them than you think it is.

I'll bet Grandpa's worm bin was in his garden; he was an efficient old feller after all, right? But I also bet it was around his tomato plants or some other area surrounded by luscious green life. I say that because your tomatoes will grow larger (if you are doing it right, and I bet he did) as the days get hotter. Larger tomatoes make shade for the worms. Voila! Your bin is out in the middle of the garden and all is still right with their world because they are in the cool shade.

You do have a couple of options with wooden boxes and their placement. You can just build a wooden box with some kind of lid that allows for air passage (you must have a lid or cover of some type, otherwise you are just making a bird buffet). A slightly propped up hinged lid is usually sufficient. A wooden lid that doesn't fit too snugly will also do just fine. No need to figure out some crazy, jerry-rigged system. You can also decide to make a bin with a bottom that is also light enough overall to not make it a horrible pain to move, then you can place it in the sun or shade as the seasons require (do not ever place a *plastic* bin in the sun or outside in the cold of winter). Or you can plan ahead where your tomatoes will be and make a bottomless bin. Dig it in so about 6-8 inches of the bin is underground and that will help to give your worms a protected spot (cool in summer, insulated in winter).

If your garden soil is particularly luscious and rich in organic matter (because, for instance, you add tons of manure every year—more on this later, too), you will want to put something down on the bottom and sides of the area you dug in so your worms will stay where you can get at them and their poop, otherwise they'll head out in all directions. You can bury a box with a bottom in it, but bear in mind that the constant contact with the moist soil underneath will shorten the life of the wood. (If you want to put holes in either of these bins for air, read the bug section first.)

Have zero carpentry skills? You can still have an outdoor bin. The conditions listed above relative to the heat and cold will still apply for any alternative you may come up with. You *can* use a *buried* plastic bin or similar container outside, but the thin sides and utter failure of most plastics to properly insulate mean that most non-wood bins need to be brought inside during the winter when it is freezing cold out (Or in the boiling heat of summer, if you live in one of those places where your state population doubles in the wintertime.). I have a great friend who keeps her worms in a Styrofoam cooler. They live outside in an always-shady spot of her yard most of the year. The rest of the time they are banished to the basement. Her worms are very happy because she pays attention to weather conditions. Once you get the opportunity to use worm castings on your plants and see the results, you will find it easier to remember to pay attention to your worms.

You can also use unconventional items as a home to raise your worms in. "Think outside the box" should be your mantra in this area. You can use discarded refrigerators and freezers laid on their backs (doors propped open a bit), old bathtubs, cabinetry, even an old toilet or old tires can house your worms. There are not too many limits other than you need something that is both sturdy and that will allow for airflow and moisture retention and protect your worms from the weather extremes of whatever region you live in.

I will, however, give you this caveat: before you attempt to grow worms outdoors, try to grow them inside for at least three months. "Inside" gives you more education about what your worms need and like than "outside" will. Sometimes, "outside" also means "out-of-mind". Growing them indoors for a few months will give you the opportunity to learn what worms are like and get used to their habits and needs. And by the time two months goes by, you will be able to see the "soil" they are making and, like any good gardener or recycler, you will start to get perhaps a little too excited about the results.

This will help you remember to take good care of them if you want to start some outdoors even though your worms will not be right where you can see them.

Windrows

Worms outside can also be kept in a trench or pit or in what is called a windrow. A windrow (pronounced just like the words that make it up) is basically just a pile of worms and bedding that is longer than it is wide and no more than three feet high. Yep, you can also just make a pile. To a certain extent. Windrows first, then trenches and pits...

Like I said, a windrow is basically just a long pile, longer than it is wide but no more than three feet deep. What?! Didn't I say earlier that your bin bedding should not be more than a mere eight inches deep? Why yes, yes I did. And no, this is not a contradiction. Conditions in an *indoor* bin are more limited than they are *outdoors*, period. Make sense? If so, read on. If not, I don't know how to possibly explain the difference between outdoors and indoors, darn it (wink wink).

Windrows are generally placed directly on top of the ground. The ground can be either actual ground or prepared ground or cement or a similar hard surface. They can also be dug a few inches into the ground. How far down is open for debate since at some point a dug-in windrow is actually a trench. I will let you decide exactly how deep a windrow can be before it is a trench.

When I say "prepared ground", what I mean is that you may want to place something between the ground and your worms. Where I live, we have moles, voles, and pocket gophers. Depending on which dastardly beast it is, they eat the worm food and even the worms themselves. As such, I place a double layer of chicken wire and some rubber backed carpet on top of the ground to help protect my worms from becoming dinner (the carpet is to help make a smooth surface to later harvest the worm poop with a shovel, it won't stop a varmint by itself, the little buggers!). If you have worm-eatin' critters on your property, or rather, under your property, you will want to give consideration to using prepared ground or cement under your worms.

I do have some windrows on some areas of pretty tough ground that is not prepared. (We have a lovely mix of clay and sand here in Colorado, which is pretty much the exact same thing as putting them on cement.) People who take my classes always ask: don't they just go into the ground and go away? The answer is: they do and they don't.

They certainly can decide to strike out on their own if you have very fertile soil with lots of yummy decomposing organic matter for them to eat. But they also aren't stupid. Okay, well they are stupid. At least according to every I.Q. test I've given them...Let's just say that they know what side their bread is buttered on and if you keep their bedding moist and keep 'em rolling in chow, they'll stay at the party. A few party poopers will probably leave, or die trying. The vast majority will stay right where you want them: around the food in your windrow. Don't your guests at a party always hang around in the kitchen? Worms would be excellent party guests in this regard, but they are horrible conversationalists and as such not the best party guests after all.

Over all it is great to put them on your least improved area of ground and then wait for them to work their magic on the bedding and food...At the same time, they will turn that least improved area into the best plot of land you own due to the poop they are making and the leachate seeping into the soil. We'll talk about what goes into the windrow when we talk about bin bedding and worm food.

Because the worms need to stay moist in order to breathe (That sliminess that makes proper little girls the world over scream "EWWWW!"? Worms are slimy because they breathe through their skin, not through lungs like us. They are able to have the needed air exchange with their environment as long as 1) there is oxygen for them and 2) they have appropriate moisture to facilitate this exchange. Appropriate = not too little OR too much.), the hardest part about a windrow, outside of harvesting, is making sure they stay moist even though they are living in the great outdoors. No, you cannot cover them with plastic to keep the moisture in.

28

Oxygen does not travel freely through plastic. You can cover them with anything that allows adequate air exchange. I cover mine with old used carpet obtained for free via Craig's List and Freecycle. (There is no such thing as a shortage of free yucky carpet, trust me on this. If you advertise for it, make sure you specify exactly how little you need or you will get an entire truckload of used carpet from a renovated ski condo complex. Just as a "for instance".)

In the winter, I cover them first with old carpet pad- but not entirely, I leave about a half foot around the edges for air to get through and a few larger gaps in between pieces on the top if they are more than four feet long or so; then cover the whole shebang with 3-4 layers of carpet, weather dependent. The pad helps hold in moisture but it also doesn't let in much air so you don't want to cover everything with it. Usually two layers are sufficient for our winters: we get a lot of snow, but also a lot of sunshine. In the summer I just use one layer of carpet and a smaller layer of pad. I know someone who covers his windrows with a couple of layers of carpet and "tents" of corrugated fiberglass greenhouse panels over the top for heat in the winter. Works quite well. A thick layer of straw or old hay will also work well as insulation and you can cover that with a layer of carpet to keep the wind from scattering it.

Use your imagination and preferably use something that someone else thinks is trash. It is more environmentally sound, plus, rather poetic that trash is keeping your trash-eaters alive. A few cinder block pieces or heavy pavers or even old tires will weigh down the ends enough to keep the wind from exposing the windrow and drying it out. Your neighbors might not like it, but if you live in a covenant-controlled neighborhood, you'll never get permission to do this anyway. No worries, you have other options. No one need ever see your outside worms if you do it right.

Worried about the chemicals in carpet? By the time the carpet gets to you, these chemicals have off-gassed sufficiently to not be a consideration. Eventually, the carpet will disintegrate as the worms will chomp on the adhesive on the underside at every opportunity. Most carpet these days is made from non-organic material. This means that you will have a few specks of carpet fiber in your worm compost most of the time. It won't hurt your plants, promise. No matter what, it is still better than putting the carpet in the landfill before it has been used for every possible purpose. If you are truly concerned about it, make sure you only use wool carpet or some other completely biodegradable covering. I do like to put a layer or two, weather dependent, of corrugated cardboard between the windrow and the carpet. Worms totally dig corrugated cardboard. They go into the little corrugated part and treat it like a love shack. They also chomp the heck out of the adhesive, then the cardboard itself as it gets moist. Good thing cardboard is easy to come by.

You have two options for the moisture component of a windrow or similarly large endeavor: you can water by hand with a hose or you can have a life outside of taking care of your worms and you can embed a soaker hose or two in the bedding as you build your windrow and just attach the hose for an hour or so every week as needed. I don't have a life outside of the farm, but choose the soaker hose option anyway due to my inherent laziness. You never know when you could get a life and you might appreciate the extra time savings.

Placement is less of a consideration with windrows than it is with other types of worm homes. A windrow directly on the worst area of ground you have water access to will turn it into the greatest area of ground inside of a year. Good ground regardless in less than a season. If you have wicked hot summers or freezing winters, direction of the windrow in relation to the sun will be a factor for you. Place them so the long side is more exposed to the sun in the winter (facing or horizontal to the south and west in the northern hemisphere) and with the short sides more exposed to the sun in the summer. This will help to mitigate the effects of the heat and use it to your advantage when you need it most.

Trenches and Pits

You can dig yourself a nice pit or trench, not more than about two feet deep, to house your worms. There are some problems with trenches and pits, not the least of which is the fact that you have to *dig* them. Drag. Depending on the surrounding soil, you will want to protect the sides from a mass exodus to the yummy, fertile soil outside the bounds of your laboriously dug trench. Another problem is that you can't move them around with ease. I'm lazy by nature. Digging a pit every year or two for my worms gives me the heebie-jeebies. But if you really, really, really want to keep your worms outside and you really only have one spot to do it in, a trench is a great option. This is also a great option if you live in one of those covenant-controlled neighborhoods since you can dig a trench (in the dark, of course, so the nosy neighbors don't report you) and camouflage the covering and no one need ever know it is there.

Although by definition a pit and a trench are not the same thing, I will use the terms interchangeably. The conditions are the same as far as the worms are concerned. Regardless of what you are digging, you really don't want to make it more than about two feet deep, max. Unless you are young and strong, in which case you can go a whole two and a half feet. Remember, digging the trench in is only half the fun, you will still have to get in there to dig the finished compost *out*.

You will definitely want some kind of cover to prevent a horde of robins living in your yard (unless you are an Audubon Society charter member, in which case: you're welcome. Now you have a cool way to attract every single bird within five miles.). Again, a layer or two of old, used carpet is sufficient. Don't forget to weigh the carpet down, just like you would with a windrow. If you do live somewhere that this set up will cause a neighborly rift, you can improvise with some carpet made of fake grass (i.e.: like something that rhymes with Blasto Smurf) and use ceramic planters with flowers in them to weigh down the edges. This will camouflage your trench so the neighborhood busy body doesn't turn you in to the all-seeing, all-knowing, all-silly-heads, H.O.A. board.

Depending on your circumstances and desires, you may want to line your pit sides with boards or thick plastic sheeting. Boards will eventually disintegrate (in a couple or four years, depending on their thickness) because your worms need to be kept moist, but are a must if you have moles or the like in your area. Plastic will also eventually lose its integrity, but if you really prefer this option, use 4ml plastic at a minimum, or go to the flea market or Craig's List and buy some rubber backed carpet to line the pit with. Use your imagination to find a good material you can use and not pay much for. Just remember that it *must* be somewhat durable. If you fudge this step and your liner falls apart, you get to dig *everything* out and re-do it. Then put everything back. A little diligence at the start allows you to revel in your laziness later.

Trenches are another great thing to use a soaker hose on. One of the major advantages of an in-ground worm home is that the ground will serve as an insulator and also help with moisture retention. You will use less water for a pit or trench than you will with a windrow or outdoor bin. But you will also have a much better chance of throwing your back out when you make the trench and when you harvest. Plus, they are not easily moved if you decide you don't like where they are or want to put the new deck in the same spot.

The methods of feeding and caring for your worms are the same for a pit or trench or windrow or outdoor bin. (We are getting to the specifics of feeding and care here pretty quick, so hold on.) The decision on which to use will largely be based on your environment. If you live on several acres and have a ready water source, you can trench or place a windrow just about anywhere. If you have a smaller yard and a shady spot, a small outdoor bin is likely the best bet if you cannot grow the worms inside for some reason. Just make sure you start with just one indoor bin or two. That experience will help you decide if you are ready for an outdoor home and give you plenty of time to plan the specifics.

If you keep backyard chickens, make sure you keep your outside worm home in an area where they cannot get to it. Chickens eat worms. Worms can't eat your refuse and give you wonderful worm poop when they are eaten. 'Nuff said. (Okay, almost 'nuff said: giving your chickens a half hour every three or four weeks to pick through your trench or windrow will not hurt your compost-making efforts overall and will benefit your chickens through increased health and vitality. An outdoor bin or box, however, will be too small for this so just remove a handful or so of the "extra" worms in your bin every month or three and toss it to the chickens. They will know what to do.)

Where NOT to Put Your Worms

You can't put worms in your compost pile. Well, pretty much. I get calls all the time, I mean ALL THE TIME from people who want to buy worms to put them in their compost pile. Or worse: their compost bin. Also their garden, their yard, their horse poo pile...All are not the best ideas. Of any, the yard or the horse poo pile are the best of four non-ideal options.

Composting is not rocket science, but there is actual science behind it. (If you want to know more, a lot more, start studying soil science. Making compost properly is an offshoot of soil science. Didn't know there was such a thing as soil science? Oh, brother, are you in for a treat! We know more about the human brain and its functions than we do about soil. Seriously. If you like science and like "green", become a soil scientist. We *need* more of you.) If you are doing it right, home composting, and by this I *don't* mean home vermicomposting, is generally a hot process. Temperatures will go past the boiling point for water. Especially in a compost bin or turner. Worms like the same temperatures you like. Ever stick your hand in a big pile of composting grass clippings? Ouch! Your worms will fry quicker than an egg on the sidewalk on a summer day in Tulsa. No, I don't want to hear about how there are worms in your compost pile. If they are anywhere but at the bottom *and* you maintain your pile properly, I'll eat my hat.

Not that I wear hats. But still...If they are in various areas besides the bottom, you are either not watering and turning your pile properly or your compost pile is "finished".

Composting properly requires that you 1) mix your organic ingredients both well *and* more than once or twice, 2) have a good mix of carbon and nitrogen sources, 3) have sufficient oxygen, 4) have sufficient moisture and 5) have temperatures that encourage *thermophilic* microorganisms. Thermophilic means heat loving. Like: well over 100 degrees heat loving. Most people skip huge on the mixing part and, by extension, the oxygen part (if you don't mix regularly, your compost pile becomes compacted and therefore air -oxygen- is greatly reduced). You have to keep mixing at regular intervals to get the needed oxygen into the mix (and to keep the icky smells away). Composting properly gives you a) no time for a social life and b) heat which would kill your worms.

The reason people think worms live in there anyway is because when the hot part of the process goes away (or never comes because you have better things to do than water and turn your pile), they will come from the surrounding natural environment to eat the goodies they think you left there for them. Most everyone composts on the ground. Um, hello!?! That's where worms live naturally. That's where they come from. This is why you think they are living in the compost naturally. But when it rains really hard or when you remember to turn and water your pile, they go away when the pile starts heating up again. Or die trying.

The fact that they go away is the part that is the most important. That and the fact that you may rarely ever remember to even water your compost pile, much less turn it. Putting worms in your compost pile is very much like just setting them loose wherever. The whole point of vermicomposting is to have their castings available when and where you need them. It's super great to have worms in your yard and garden, they provide infinite benefits. But just setting them free is not the best choice and that is what you are doing when you decide to put them in the compost pile (or sentencing them to die by being essentially boiled to death).

If you are not watering and turning your pile, and therefore not reaching the required level of heat, you are doing what is called "cold composting". Ever notice that is seems to take *forever* to get compost out of your pile? That's just a symptom of your laziness. Fear not! You are in good company. And vermicomposting is *exactly* the lazy process you need to save you and your garden.

Putting worms in your horse poo pile (Or cow. Or alpaca. Or sheep...Or, or, or. Most farm animal poo piles are essentially the same except for fowl poop—notice I didn't say "foul" poop! Fowl poop runs so hot –high in nitrogen like your grass clippings- it must be cut with something else, a carbon source, and pre-composted to make it suitable for worms to live in.) is not the best idea unless you are going to turn your horse poo pile into a proper windrow and make sure it is sufficiently watered. Where I live, you can't swing a Parelli halter without running into a neighbor with horses. Most, to my great chagrin, toss their horse's leavings in the dumpster to be hauled to the landfill. Some like to pile it up, let it sit for a year or seven, and then post it on Craig's List as "compost" either for sale or for free. Without proper management (broken record), sorry folks it is just aged manure, not actually compost. Not that it doesn't have its benefits to the people who flock to add it to their gardens. But let's call it what it is: old poop. Not compost.

Most farm animal manures come with some kind of bedding. Straw, uneaten hay, wood shavings, etc. The manures combined with the bedding make for a great carbon to nitrogen ratio for hot composting. Add water and a little air and voila! You are hot composting. Right until the initial heat dissipates due to lack of oxygen because the reality is that only about .002% of people with these piles actually turn and water and *manage* their manure piles (a figure I totally just made up but is absolutely very close to accurate). However, building the horse poo piles in windrow fashion and making sure they have sufficient water (with soaker hoses or the like) and air will make them great for vermicomposting. But you must take these extra steps. Otherwise it's just a pile of old poop.

So what? It's all the same, you say. No matter what process you use, it will result in the same thing, you tell me. Right? Wrong. Actually, "wrong" is just not a strong enough word. The difference between composting and cold composting and vermicomposting and a pile of old poop are crucial. They may look very similar, *but your plants know the difference.* Each process depends on microbial activity to be successful. These microbes, through many processes that are not fully understood, provide your plant with the necessary nutrients and micronutrients to grow strong and healthy. More importantly, strong and healthy plants can fend off fungi and disease and insect predators *on their own.* Did you know that plants have hormones? No, really, they do! I didn't know that before either, so again, you aren't alone. Plants also need a large variety of nutrients and micronutrients. These big and little nutrients are vital to their *true* health, not just the *appearance* of health.

To make it very simple: these microorganisms put these nutrients and micronutrients into the soil in a form that your plants can use. The "in a form your plants can use" part being crucial. It's very much like you taking fat soluble vitamins with just a glass of water. The vitamins won't be used properly because they are not presented in a way your body can use them. Same thing with plants. Cold composting or, "aging", will give you decent plant nutrients. Hot composting will give you good plant nutrients. Vermicomposting, depending on the individual nutrients that you are looking at, is at least five to ten times better, or more "nutritious", than what even hot composting will give you. You may not get as much final product with vermicomposting, but what you do get is worth so much more to your plants per pound than pretty much anything else you can possibly give them. Plus, it's way easier on your back. Super double bonus points for that.

Besides wanting worms to put in the compost pile, I also frequently get folks that want to put them in their garden or yard. Someday, I will have one of those phones that you can "press 1" for this and "press 2" for that and all there will be is "press 1 if you think you want to put worms in your compost pile" and "press 2 if you want to put worms in your garden or yard" because it's always the same answer I figure it will save me a ton o' time in the long run.

If I did do that and you did press '1' or '2', this is what you would hear: Don't do it. Just like putting worms in the compost pile is a bad idea, so is putting them directly into your yard or garden. Though not for the same reasons. Are worms super beneficial for your outdoor spots? Heck yes! Otherwise, why would I go to all the trouble of trying to keep my cat off my keyboard so I can tell you all this? Yes, worms will aerate the soil and chomp up decomposing organic waste and leave behind valuable pieces of poop that will in turn help provide vital nutrients for your grass and plants and so on and so on. This is a no-brainer. That is why everyone asks to buy some to put in their garden and yard. Yep, I get that part.

So what's the problem? The problem is that if you have to buy worms to put in your garden or yard, that means you think you have either very few already or, worse, none. If that is the case, there is *a reason for that*. That reason is why I say 'no'.

Worms, to point out the obvious, are living creatures. This means that, unlike a brick, they have to have certain things to sustain life and reproduce. If you don't have worms in your soil already, then likely you don't have those things they need to survive and thrive. Which is the reason nothing is growing: not because you don't have worms but because you don't have the necessary components to *sustain life*. At this point you have two choices: start a rock garden or get the stuff that sustains life. Notice neither of these two choices is: put some worms in your lifeless dirt. That would be a death sentence. Might as well just put them in the middle of the highway.

If you want worms because your yard or garden has such poor soil that there are few or no worms in it already then what you need to do is make the soil better first. There are books upon books and magazines and internet articles ad nauseam on this topic. Essentially: compost (and vermicompost, naturally) are your best friends in this endeavor. Worms must have air, moisture and food to survive. Your soil must therefore: be able to hold moisture, allow for air passage (meaning: not be compacted and cement-like) and provide food for the worms and other microorganisms to live there and work their magic. Reading just about any of the organic gardening or soil health books at your library or getting pretty darn near any one of them at the bookstore will set you on the right path.

If your soil and garden plants *are* growing but you have found you must continually add chemical fertilizers to make this happen and would like to try a more organic approach then my answer is the same: your soil is dead. The petroleum and chemical based products that you have been using for years (fertilizers, pesticides, herbicides, fungicides) have killed the microbial life in your soil. That is why you have to keep adding this stuff, the plants are not healthy enough to grow on their own and defend themselves because they are not getting the micronutrients and other goodies they would get from healthy soil. Your soil is not there just to prop up the plants, people! If you are one of these garden-chemical-heads, start from the same place as the person who already knows they have dead soil to rebuild your soil health from step one 'cause, sister, you **do** have dead soil, you're just missing the paperwork to prove it.

"Well", you say, "that's not me." I have been organically gardening for years and both my yard and my garden are teeming with worms. I just want MORE! This group of people I call the "take two, they're small" group. They think if a thousand worms are good then two thousand must be at least twice as good. No. Yes, more worms *are* good. But if your soil could support a higher worm population *then they would already be there.* This means that if you have the proper air, moisture and food for your worms, then they would already be reproducing *proportionately* to the environmental conditions and the space allowed. They don't keep reproducing until they take over an area, otherwise my garden would be a roiling mass of wet, pink worms instead of nice green plants.

Look at it this way: if your current household supports you, your spouse, your three kids, two dogs, four goldfish and one cat in perfect style and harmony, then adding another couple, three kids, two dogs, four goldfish and one more cat

38

would make things *twice as* harmonious, right? Of course not. You'd constantly be in each others way, there would never be enough food and I won't even start on the hot water and towel situation. Same thing with the worms. Well, except for the hot water and towel part. Make sense now?

Adding compost and other organic matter along with vermicompost will increase your soil health and worm populations **naturally**. Because worm cocoons (worm eggs) are naturally present in vermicompost, you will have a good base to start with and the regular compost will provide food for your worms and their population will grow to sustainable levels based on the conditions present in their environment.

Adding more worms will also not make your garden produce uber-fruit suitable for entry into the "Biggest Tomato in the County" contest at your local fair. Their poop will absolutely increase plant health, which will give you stronger, more resilient and disease-resistant plants which in turn will provide you with higher and better-tasting yields than you've ever had before. But those uber-fruit you dream of come about as a result of seed selection and careful pruning and such, *in addition to* proper soil health, not just because of soil health alone. A cherry tomato seed will not produce a beefsteak-sized tomato no matter how many worms you have in your garden. Worms are just one component of the whole picture; but don't get me wrong, they are a vital part.

Now that we have the whole picture of the proper real estate, let's explore the furnishings...

The Furnishings:

Proper Worm Bedding

Oh, the number of things you can use as bedding! Shredded newspaper, junk mail, cardboard, paperboard, wood shavings, coir (coconut fiber), peat moss, pre-composted manures, finished compost, straw, old hay...Just about anything organic (organic in the sense that it was once alive, not organic in the sense that it was grown without pesticides) and not too high in nitrogen will work (high nitrogen = high heat; remember that big pile of grass clippings). Let's explore the most common options for your worms and I'll save the best for last so do stay tuned:

Peat Moss

If you have ever bought worms over the internet, likely they were shipped to you with peat moss as bedding. Yes, you *can* use peat moss as bedding. However... There are a couple of good (or great) reasons not to. For one, you have to buy it. If you are trying to raise worms to recycle your kitchen scraps and other organic waste, then *buying* something to use for bedding doesn't seem to make much sense to me. Yes, yes, I know you would buy or have bought a special waste can for your organic recyclables if you're lucky enough to live in a city that has a program like that. But generally you only buy that once.

The second, and most important reason to not use peat moss as your bedding is because most of the peat moss you would have access to is going to have originally come from *non-renewable* sources. Naturally formed peat bogs do not replenish themselves overnight any more than oil deposits do. To say the least. Purchasing peat from these bogs is like voting with your dollars to say it is okay that these bogs are destroyed. It's not. They are there for a reason and provide special and very specific habitat to innumerable creatures. We won't even begin to mention the added environmental havoc caused by all the transportation of the harvested peat moss. If you want to know more, do an internet search for "sustainability of peat moss" and you'll see what I mean.

If for some reason you are still hopelessly enamored with the idea of using peat moss, you should probably soak and rinse it at least twice before you use it as bedding as it is naturally too acidic to be used as worm bedding as is. If *that* amount of added labor is not enough to dissuade you from using it, I'm at a loss.

Coir

Coir comes from coconut shells. That hairy stuff on the coconut? Coir. I've heard it pronounced both like "core" and like "noir", as in pinot or film. I don't know which and the dictionaries I've looked at aren't much help. Although this meets the "renewable" aspect, it fails utterly in the categories of "not spending money on your worms" and "not being readily available". Again, it has to be transported. Think of where you live. Now think of where the nearest coconuts would readily be in large numbers. If you can't walk there or bike there and throw some coconut fiber in your backpack for your worms, you should probably not use this. If you can, then it merely needs to be soaked, but does not need to be rinsed like peat moss. But know this: just because you live in an area where coconuts are so abundant they are dropping on people's heads left and right, does not mean that the coir you buy in your local garden shop is from your area. Read the label. Most coir sold commercially is imported.

Wood Chips and Sawdust

Most of you will have a source of itsy-bitsy pieces of wood readily available even if you don't know it yet. Sometimes you just have to put it out there that you are in the "market" and your woodworking friends will come out of...well, the woodwork! Chips are bigger than dust, but with woodworking you usually will get some combination of the two. The most important thing is that the wood chips and sawdust come from wood that has not been treated in any way. Chips or dust from painted or stained or pressure treated wood is a no-no for your worms until more research can be done to determine what sort of chemical residues are left in the finished compost and if these residues are taken up by the plants.

I've had varying success with wood as bedding. I have heard of many folks who use it in combination or as straight bedding with great success. I have access to huge amounts of free sawdust with some wood chips in it. I used it in combination with other bedding for quite awhile, until one day I got a new sawdust delivery in at the same time I was harvesting and making a ton of new bins. I threw in the sawdust just like I had been doing and added the worms as I harvested them. All exactly like I had done before. Within a week, all of the bins made with the sawdust from this new batch were dead or dying. Amazingly enough, the worms had not crawled out to escape, which is pretty normal behavior when conditions are not great for worms (we will definitely get to that part soon enough). They were just laying there in the bedding, essentially disintegrating before my very eyes.

The only thing I could think was that the sawdust was too resinous. It did smell like pine to a certain extent, but I have no idea what kind of wood made the dust, I only knew my worms were lost. Mostly. I noticed that some were happy as clams and not showing any signs of dying. I harvested the healthy ones I could find from the bins and put them in a "regular" bin with just paper, as I previously used to make all my indoor bins. I left one of the sawdust bins alone with the remaining worms that seemed healthy. Within about two weeks, about a third of those had died. The rest remained healthy and though sawdust takes a bit longer than paper to be eaten and turn into poop, turn it they did.

The moral of this story? Sawdust and woodchips are usually fine. I know for sure lots of people use them and experience no problems. I have heard that aromatic woods should not be used as sawdust or to make wooden bins. However, I also know that there are people who raise worms where cedar is prevalent and make bins from it and use cedar sawdust and have zero problems. I don't use sawdust anymore, instead I give it to people to use as bedding for their livestock, then I take it back after it is "used" and mixed with manure. I have no problems with it after it has been used.

I am not against sawdust and woodchips at all. I would, however, highly recommend to add them gradually rather than all at once to let your worms get used to them. Your experience will hopefully be much, much better than mine. As far as I'm concerned, I'll leave the sawdust to the livestock.

Straw and Hay

Lots and lots of folks have access to straw and/or hay. You may either have livestock or live in the country or know someone who does. Straw and hay are great for bedding. And they suck as bedding. Straw is slightly worse, but both have a major drawback: they don't hold moisture well. I know, I know: If you use either of these for your livestock and have had them rained on, you are going to beg to differ. And for the purpose of feeding and bedding your livestock, you are right. But for worm purposes, you are not. Straw and hay are superb in their ability to keep a bin properly aerated when used in combination with other beddings. They are, however, terrible at keeping conditions adequately moist for your worms. "Too much moisture" in a straw or hay bale meant for your livestock is often not quite enough moisture for the worms.

Straw and hay are commonly found in combination with some kind of animal poop because these are used as food and/or bedding. Rabbit, horse, cow, alpaca, etc. When it is in combination with poop, it can be super cool to use. I also have hay that has gotten wet and is not in sufficient quantity to sell as cow or sheep food. I use it with other beddings at about a quarter to a third of the total. It can help keep a bin properly aerated because the stems retain their fibrousness for quite awhile and help prop up the rest of the bedding and keep it from becoming compacted.

This is also the other problem with using straw or hay as bedding: because they are so fibrous, they can take awhile to break down and get eaten. I think using hay and straw will depend largely on their availability to you. If you have straw and hay lying around and it can't be used for something else, by all means, recycle it into amazingly valuable worm poop. I would not, however, go out of my way to get either of these unless I needed them for insulation for an outdoor worm home or the like.

Manure

Let's talk first about "vegetarian" mammal manure and then we will get to the inevitable question of dog and cat poo. (If you raise any of the following animals, please understand I say "vegetarian" for any of these animals knowing full well that they are not all herbivores but may be perceived to be. If you don't own these animals or have familiarity with them, I'll let you research for your own knowledge why I put "vegetarian" in quotes.)

Horse, cow, llama, alpaca, goat, sheep, hog and rabbit manure are the most commonly used for vermicomposting. Any of the ruminants listed will give you a better starting product than the non-ruminants. Meaning: if the food has to go through more than one stomach, as it does in a ruminant animal, it will be more digested at the end, too. More digested = more easily converted by your worms. Additionally, if the poop is naturally or usually collected with the bedding (if any) or leftover food (usually hay or tough-shelled grain), it will take longer to be converted into worm poop.

Chicken, turkey, duck, goose (duck, duck GOOSE! You're it.) and other fowl poop are the second most common manures that people have access to and want to use in their worm bins. I say YES! And...NO! Fowl poop is very foul for your worms if it is not **pre-composted**. It is so high in nitrogen that it falls under the same rule as green grass clippings: it gets hot, hot, hot. So my NO is for non-pre-composted fowl poo and the YES is for pre-composted fowl poo.

Rabbit manure is also very commonly used for worm food. More people than you know have more rabbits than you know and therefore, much more rabbit poop than you *want* to know about. I know there is a small book written specifically regarding commercial rabbit production with worms housed right under the pens and I'm told this works wonderfully. I don't currently raise rabbits, but do have people who bring me rabbit poop mixed with bedding (yes, I am easy to shop for at Christmas) and I treat it like any other large animal manure.

All that said, animal manures are GREAT. They are actually generally useful as either food or bedding or both. Convenient, eh? But like all good things, there are some crucial rules that you don't get to make exceptions for (trust me, the rules will prove themselves).

Rule #1: if you give your animals any kind of deworming medication, you absolutely *must* make sure that you do not put the poop from your recently dewormed animals in your worm bin until at *least* two weeks have passed. Two months is better. During that time, you should expose the poop to either plenty of rain or plenty of water from a hose to help it along. Remember too that getting rained on doesn't count if your pile has formed a "crust" at the top. The rain will roll right off of that. I like to stick a hose down into the pile and let it run for awhile to help things along. Works great.

Rule #2: Urine is generally high in salt. Salt is one of those few organic materials that is both detrimental to plant life in high quantities and not converted by worms into something beneficial. We'll cover salt more when we talk about food, but understand that salt is not your friend in this case. Yep, it's yummy on food and yep, it'll kill your worms if you give it to them in high quantities and kill your plants if it is not leached sufficiently before you use it. This is mostly only going to matter to you if the animals you collect the manure from are confined in a manner that causes them to urinate and defecate in the same place. Pre-composting and/or leaching the manure with water will help reduce the salt content.

Depending on the amount of manure you are dealing with, it is vital that you take potential run-off into consideration when leaching the poo for medications and salt. Potential run-off must be monitored and addressed lest your recycling efforts cause more problems than they solve.

Rule #3: Most but not all of the poop you collect from your mammals and birds will be "hot" at first and needs to be at least slightly pre-composted or, at a minimum, aged somewhat before you use it. A month is usually more than sufficient for most non-fowl manures.

Okay, Ms. Rules (um, I prefer Mizz Rulezzz, if ya don't mind), you say, why would I want to go to all this trouble to feed this to my worms when I can put it right on my rose bushes as is? Well, I'll tell ya...Yep, putting some poop on your roses or in your garden is usually great. However, the *value* to your plants of regular poop vs. poop that was first eaten by worms is like the difference between one of those mini-chocolate bars you steal from your kids at Halloween and having a personal tour of the Godiva Chocolate™ factory. Plus, worm poop looks nicer and smells waaaaaaaaaay better. Plus plus, it is unlikely that you are having as much success as you could be having growing green things if you are really using fresh poop as a fertilizer.

Most folks leave their animal poop in a big ol' pile for a few months until they can get around to using it or disposing of it (life on a farm keeps ya busy!). Some people even try to sell or give away this result as "compost". In reality, it is merely old poo. Not that it doesn't have great value as organic matter (we'll get to that more later, it's important) and some value as fertilizer for your plants. But it is not "compost".

You folks that do have these piles are also going to try to dispute the above listed rules. You'll go on and on about how whenever you do move your pile, there are (all together now) "tons of worms in there!". Look here, people, they don't call me Mizz Rulezzz for nuthin'. If you will recall, or if you will just put down this book for a second (I know it's hard to do, but it's only for a little bit.), you will observe that the worms are really only at the *bottom* of your pile. Same thing regardless of the fowl or mammal question, the worms are not all over the pile. Okay, you with me? Now here's why:

Remember I just said that these piles can be hot? Well, the bottom of any given manure pile is the "first in" stuff. It's been sitting there for awhile since we don't just make pretty piles of poop and then get rid of them promptly. No, we make pretty piles of poop and then, because "poop happens" and I do mean *daily*, we pile new poop on top of old poop. (Usually exactly the height of either your tractor loader or however high you can fling one shovel-full of fresh poop.)

Gravity sucks. It acts on all things, including moisture. The moisture naturally present in the fresh manure and/or from rainfall is subject to the same laws of gravity that are responsible for my skinned knees. ("Grace" is so not my middle name.) Your haphazard pile-making skills add air to the pile. Moisture in fresh poop is a given. These things act on the lower parts of the pile to start making hot compost. You put more poo on the pile and soon enough, things go anaerobic because the weight of the new poop is compressing the air pockets you put in when you were making the pile. That's why you balk at trying to use the poop pile "too soon". It stinks to high heaven in the middle and at the bottom. Enough air will generally get to the top and sides to keep those areas from stinking too badly, but they also dry out sooner so generally don't break down at the same rate.

But! After awhile of this, the wee little anaerobic bacteria that were stinking things up, "climb up" and into the pile where things are more to their liking and fresh, hot poop is being piled. And that is when the worms living in your soil or their cocoons that have been left to hang around for just the right conditions to hatch decide to make their new home. They will start near enough to the bottom edges where there is more room and work their way in as the stinky stuff goes upwards.

One of the insanely cool parts about worms is that they naturally aerate the soil. Their movements add air channels, but their deposits (castings/worm poop) are also fluffier. The more air channels they make with their movements and fluffy worm poop, the farther in they can go because there is more air. More air means they can live further in the pile to get to more food, regardless of how high you pile it. The worms will keep doing what they do best, eating your animals decomposing waste, as long as conditions are good. Which is why they are not all through the pile: the conditions are not good for them in the stuff up higher that is either too fresh, not wet enough or too anaerobic.

So, yes, you little rulezzz-breaker, there are worms in your manure pile. But no, sadly, they are not going to transform the whole pile the way you have it now. Unless...

My friend Jack is a hoot. You should meet him, really. A retired city fire-fighter and long time cowboy, Jack still keeps his hand in and raises a few head of beeves (that's the plural form of the cattle raised for meat for all you city folk) on his acreage a couple exits away from mine. Jack is contrary by nature, which is likely why I'm so fond of him. He would beg to differ with my statements about worms not being all through his manure pile. And he's right. What makes the difference in his case is that his poop pile is at the outside corner of his barn. The barn whose roof just so happens to direct the rainfall right onto, you guessed it: the top of the pile. This rainfall makes awesome worm conditions in the pile. The water falls in about the same spot whenever it rains even a tiny bit. This introduces both water *and* air (because of the channel the water makes, not because there is oxygen in water -H_2O-

which we'll cover later) into the middle of the pile. Except for the very outer bits, his pile is teaming with worms and is incredibly quickly transformed from a huge pile of steer poop into a huge pile of worm poop. His wife refuses to let me have the contents of the pile. She's just that way. You want to see *green* in an area known as "dryland" pasture (emphasis on the dry) ? Go to Jack's house.

Livestock manure is a great thing to use for your worms if you have ready access to it and a superb way to recycle your animal manures on your farm to use in your pasture and replenish your soil. I would not go out of my way to drive into the country to fill my car trunk with manure to feed my worms, but if you already have it, vermicomposting is the best possible thing you can do with your animal manures. Period.

Okay, okay already, I'll talk about composting dog and cat poop. Props to all you city-dwellers who made it through the livestock manure section. Way to hang! One of the most commonly asked questions I get is "Can I put dog poop/cat poop in my worm bin?". The answer is...Yes. And no. (Yes, I know I keep doing that, but not everything is black and white in the worm world. Don't yell at me, I didn't invent worms.)

Here's the gist of the NO part first: dogs and cats have some nasties in their guts and some of those nasties survive in their poops. E. coli and salmonella and listeria and the like, not to mention toxoplasmosis. Plus, you still have a bit of the "hot composting" issue that you can have with other manures because of the undigested grains in the poo. But the nasties are the main problem.

There have been studies (Yep, plural. Meaning: not just anecdotal evidence.) where these nasty bacteria have been planted on purpose into worm windrows, measured, allowed to go through the worm eating and pooping process and then measured again and hmmmm, the levels are so drastically reduced as to be negligible in comparison with when they started. What does that mean? Well, that depends on a lot of things.

I'll try to not go into a long diatribe about all the food contamination scares of the last few years and their connection to industrialized agriculture and the centralization of food distribution (Arrrrrrrrrrrrrrrrrgh! Sorry, that was just a pain I got from keeping off the soapbox. Buy local and sustainable, please!) I can tell you for a fact that none of these food scares came about from anyone applying vermicompost to the food they were growing. Not even from vermicomposted dog or cat poop.

That said, some food plants will suck up whatever is in their growing medium (your dirt), not just the good stuff. There are naturally certain amounts of the nasties in lots of different soils and manures; you won't know they are there unless you test for them because most people are healthy enough that these nasties don't lay you low. And if you did test and got a positive test, that is not necessarily a reason to freak out. There is evidence that the food scares involving unprocessed food (like spinach and tomatoes, two of my favorite things) involved liquefied unfinished compost applied on the foliage (on the leaves, not specifically into or on the soil). This unfinished compost got into people from food that was not grown properly (with finished compost as opposed to raw or unfinished manures) and/or not washed properly. At least, that is what a lot of the evidence shows and what a lot of folks who are smarter than me believe is what happened. The processed food scares, like peanut butter, likely came from improperly maintained and improperly cleaned facilities. What all this means to you: if you use your pets poop to feed your worms, you are taking your chances. I'm pretty close to saying "Yes, do it!", but only pretty close because I'm not entirely convinced that 100% of the evidence is in for this particular question. That and because if I'm wrong, you could get really, really sick and I'd never want that.

But! (There she goes with 'the buts' again.) There is absolutely nothing that says you shouldn't feed this to your worms and then use the resulting worm poop on your NON-food bearing plants and trees. Worms migrate, poop doesn't (I said *migrate*, I didn't say anything about rolling downhill, smart aleck). Any nasties that might theoretically be in the poop are going to stay where you put them so please, by all means, start eating spinach and tomatoes again!

So what to do if you decide this is an option you want to explore? Let's start with dog poo since that is easier.

Unless your dogs are in a situation where their urine and poop is somehow mixed together (if they are in a dirt or grass yard where they generally perform these functions in different areas, even if only by a few inches it's okay), composting dog poop is pretty much exactly like it is for horse poop: use it as food though, not bedding (dog poo doesn't have the hay or bedding bits mixed in like horse poop does) and make sure it is at least slightly aged or, better, hot composted before you use it. Just like with livestock, if your dogs are on a deworming medicine, leach it well and let the poop age or compost it for at least two weeks before you use it. If for some reason your dogs are in a situation where their urine and poop are mixed up together, wait until that veterinary problem is resolved before you use the poop or stop whatever you are doing and let the dogs run in the yard already. They need exercise.

Cats (and I will include litter-trained ferrets in this category), because so many live inside and use a litter box, are a whole other story. If your cats go outside and you pick up their poop just as you would for a dog's poop, then you can treat it just like dog poop for vermicomposting purposes.

There is not a lot of information available regarding composting the contents of your cats litter box. As of this writing, I have three cats who live both inside and outside at their own liberty. I do not currently compost their poop. Someday maybe. What is "for sure" is that if you want to compost the litter box contents then you must not use clay litters or "clumping" litters. There are several types of cat litters made out of organic materials available at any pet store or large grocery store. Generally these are more expensive, but if you are paying to have waste hauled away, then using and composting the organic litters will commensurately reduce your waste costs (both packaging it and having it hauled away) if you are composting the litter box contents so please keep this in mind when you are gulping hard at the cost of compostable litter box materials.

That said, the "natural" litters made of recyclable materials (generally wheat chaff, pelleted newspapers, shredded corn cobs and the like) will serve pretty well as bedding material in their own right. Cat urine is high in ammonia and high enough in natural salts to make it potentially detrimental to your worms. There are people who think rinsing the used litter before adding it to the worm bin is a good idea. I agree. And that is why I don't compost my own cats litter. In a word: ewwww. But some people have stronger constitutions in this area than I apparently do and if that is you:

I'm not entirely convinced that rinsing is the only option. If you have two litter boxes you might be able to use fresh air and sunshine instead. Depending on your number of cats and the days in between cleanings, I think you could have success if you take the used litter box and put in an area where it can be exposed to fresh air and sunshine, stirring every now and again to incorporate fresh air, it is likely that the ammonia will dissipate enough to make it a negligible concern for vermicomposting. Sunshine and fresh air won't change the salt levels though, so maybe a rinse and *then* some sunshine is the right ticket.

I imagine that hot composting the contents, either alone or with other ingredients as in a regular compost heap, might also do the trick. But again, I don't know. This would be a cool group of experiments to conduct for the person with lots of cats and loads of extra time. (By that I mean: not me.) I'd love to hear about it though if this is something you have had success at or are going to experiment with. I would definitely recommend that you get the resulting vermicompost tested at a qualified laboratory for pathogen levels before you consider using it on any food crop. I would also recommend against doing this or using the resulting vermicompost with bare hands if you are pregnant. Cats can carry toxoplasmosis in their feces. Although there are studies that say the worms can mitigate this nasty, I'm not confident enough in this process to recommend you take the risk if you are pregnant or planning to be.

Okay, okay, fine. I know you are thinking it so I'll talk about it VERY briefly. Yes, you can vermicompost human poop. There have been whole other books and parts of books and articles and, Lord have mercy, even videos on YouTube and the like covering how to compost your very own, very personal waste. I will not be covering that any more in this book. Suffice to say, if you are interested in this I would treat it as I would dog or cat poop for the purposes of vermicomposting. If you want to know more, google it. "Humanure" is a keyword you will want to try. (No, I didn't just make that up.)

Leaves and Grass Clippings

I think we covered the grass clippings issue sufficiently already, but just in case you are one of those people who just flips through a book looking for the stuff you think is relevant, we'll cover it a little bit again (and also take this opportunity to tell you to go back to page one and READ THE WHOLE DARN BOOK, WILL YA? See what happens when you skip only to the sections you "need" to read? You miss important stuff tucked in other, less obvious, places. Sheesh.)

Fresh grass clippings are very high in nitrogen. High nitrogen generates lots of heat when it is initially decomposing. I mean lots of heat in the way that would basically boil your worms to death. Using green grass as bedding is a bad idea. It is a horrible idea if you are bin composting. If you are composting outside in a large trench or windrow, then green grass can be used in larger quantities, but you still want to make sure there are plenty of areas without any green grass so at least the mature worms can escape to a cool spot when the clippings go tropical.

I've read that this same crazy hot heat can be an advantage to outdoor composting in colder climates specifically because of this heat. The idea is that you can apply a somewhat heavy layer of grass clippings (mixed with something a little more carbonaceous like leaves or straw to keep them from matting and inhibiting air flow) on the top of your outdoor piles and the heat generated from the decomposing green grass will help keep things warm for your worms.

This is something else I've never tried, but think would be a good idea if logic didn't rear its ugly head every time I try to think of how to make this work. Granted, I've only lived in the West and the Midwest of the United States so maybe where you live conditions might be different. But where I live, when I need this kind of warmth to fend off the chilly or downright freezing winter temperatures is in the actual winter. I do have a landscaper who brings me the grass clippings he collects from the clients who aren't savvy enough to leave them where they lie. Try as I might, he refuses to bring me any fresh green grass in winter. I'd say it's just because he's being contrary, but he tells me he isn't. So essentially I can't get my hands on fresh grass to test this neat-o theory, but maybe where you live you have several warm days in a row where you can get out and mow the lawn followed by a week or two of below freezing where you can test the effects of a grass clippings blanket. Lemme know how that works out for you, will ya?

For the rest of the world: I'd tell you how a handful or two, even in a bin, is not going to be the end of the world. But when I try to say things like that, inevitably I get the phone call from the guy with a shoebox sized bin and hands the size of dinner plates who tells me he did "exactly what I said" and now his worms are cooked. So, I'm not going to say that.

Instead, if for some reason you don't want to leave your grass clippings on the lawn where they will do the most good (by giving valuable nitrogen back to your grass and adding decomposing matter which will be food for the worms and other beneficial critters living in your yard), then put the clippings in the hot compost pile and you can feed them to the worms after they are done composting or you can use the clippings directly as garden mulch. If you are not familiar with the benefits of just leaving the clippings or have fallen hook, line and sinker -ha, a worm joke!- into the myth that doing so will cause endless problems with thatch buildup that will kill your lawn and make you the laughingstock of the suburban lawn mower set, then please take a few moments (now is fine, I'll wait...) to look this up on the internet or call your county extension office or local master gardener to learn the truth about lawn clippings.

I love autumn. The sights, the smells, the opportunity to get things ready for the winter when I can nap on the couch in front of the wood stove. Ahhhh. But I love it most because no one ever seems to know what to do with their fallen leaves. Besides the worm classes I teach in early spring and summer, I get to do most of my edumacatin' in fall.

Because I'm known to accept certain organic waste products for recycling at my farm, I very frequently get requests to accept leaf waste in the fall. While I LOVE to get these dozens upon dozens of bags of leaves and truck beds of leaves and once, even a car trunk full of leaves, there are better things to do with them. First, if you have a worm bin then by all means you can use some of these leaves as bedding. Read the section on bugs first, but, yes, you can totally use them. It helps to crunch 'em up really well first, as the smaller they are the faster they become worm food and the faster the leaves become worm food...the faster they become worm poop! Likely, though, you will not have enough bins to use up even one whole bag of leaves unless you are or are becoming the crazy worm lady in your neighborhood. (Git your own neighborhood cuz' 'round here **I** am the crazy worm lady and there's only room in this town for one of us...so drop those worms and back away real slow like...)

If you have a yard, the best thing to do is go over them a couple of times with the mower and water them in for a wee short time and leave them there to rot where they lie. You will hear from some landscape folks that the leaves don't provide much in the way of nutritive value to your grass so you shouldn't do this. They are right about the nutrient value. Dead leaves are mostly just, well, fiber (though I still wouldn't eat them for breakfast). What these landscape people are missing is that the leaves aren't put there for the grass, they are put there as *food for the things that feed the grass*. See, your grass could take or leave the leaves (ha! the hits just keep on coming), but the worms and other nice bugs and microorganisms LOVE your leaves. Breaking them down first with the lawnmower or mulcher just helps things go faster and makes your neighbors happier; plus, it helps keep these valuable little worm size lunch plates right where you want them, instead of blowing away so they can decompose and feed the worms in your neighbors' yard. It's okay to be selfish in this area, I promise.

If you have a compost pile, they can certainly go in there. Some folks have what they feel are "too many" leaves for just the yard. If you have a garden, you can cover the whole garden in leaves, give the mess a little water to get them matted down just a bit so they don't blow away, and by spring you will have a nice mulchy layer of what used to be leaves that your plants will love (and ready worm food for the worms already living there). When I lived in town, I always liked to mow them once in the yard without the bagger then go over them a second time with the bagger attached and put the bagged leaf pieces in the garden and the rest could stay on the lawn as free fertilizer.

If you have outdoor windrows or a trench or the like, leaves can be poured in "as is" with a few well-timed water sprinklings to help in the decomposition process without saturating the worms or matting the leaves so much you cut off all air; though this is hard to do with leaves. (We will be getting to proper moisture levels pretty quickly, hang in there.)

If for some reason you find any of these ideas objectionable and just can't wrap your head around another way to sustainably dispose of them, bag 'em up and put an ad on Craigslist or Freecycle or the like and almost always someone will be inordinately happy to take them off your hands. Heck, most of the time you can find "wanted" ads for your leaves on either of these handy, free, internet exchange sites. Fort Collins, Colorado, a college town north of me, has a "Leaf Exchange" service right on their city website so the "need leaves" people can readily hook up with the "have leaves" people. How cool is that? Maybe you could start this in your area, too. If all else fails, find someone near you like me or take your leaves in for recycling. Lots of municipalities offer leaf recycling to residents.

Like grass clippings, leaves should never ever never ever never be put in a landfill. Ever. Throwing either away is like lighting your money on fire and telling the grandkids to go to heck. They are quite valuable to all things green with very little work (less work than bagging them and dragging them to the curb!) and a landfill is the last place you want to put green waste. Even if you are not ever going to have a windrow or trench situation for your worms, leaves and grass are great free mulches for your garden.

Here is the part where I get to hear from my students all of their concerns about accepting yard waste from a complete stranger. "What if they spray?" is usually the most common worry. Well, what if they do? If you are not a certified organic farm, this is not going to be a huge concern for you. Yes, I'd also rather live in a world where people wouldn't mind a few weeds in exchange for never using dangerous chemicals on their lawns. But here's the cool thing: grass clippings will quickly get hot in a confined space. "A confined space" is generally what they will be in when they come to you, be it a garbage bag or in a big pile in the back of a truck, like I get every week in the summer. What makes this cool is that the heat will help mitigate the effects of most of the chemicals that might be used, the vast majority of which are fertilizers anyway, not pesticides or fungicides. You can also be pretty darn sure there will not be much in the way of herbicides after the initial application of pre-emergents in the spring. If this is a huge worry for you, just *talk* with the people who might be providing you with some lawn clippings and find out what they use, if anything. It won't take you many conversations to find someone who uses nothing but a lawn mower. As for leaves, there are very few sprays that are applied in the fall when leaves are falling so likely these will be 'clean'. Personally, I think keeping these things out of a landfill should also be taken into consideration when deciding what's most important. Still not convinced you want to use them? Then please at least consider petitioning your local government to start a hot composting program for yard wastes; that way, we all win.

Since we are in the general area of the yard...If you have twigs and sticks and such that need disposed of, you can certainly put a few in for your worms. They will, unfortunately, take forever and a day to break down in a regular bin or windrow situation. Though a few twigs can really help keep things properly aerated so by all means, add some. Twigs and sticks are frequently tossed with the trash and every time I see them out at curbside I want to stop the car and rescue them (and us) from a useless landfill death. If you have a fireplace or know someone who does, use these as kindling. If not, back to the internet to list them for someone who does. Once you find someone to take them, make sure you get their information so you can call them in the future when you have more. Very little effort is required to recycle all sorts of bulk organic waste like this, you just have to have a tiny bit of patience and perseverance to find the closest person who needs it and ~voila!~ a great symbiotic relationship is born and over time, you can literally save TONS of organic debris from going to a landfill. Tons. Think about it.

Speaking of fireplaces: Yes, you can certainly put some of the ashes in your worm bin. They will get eaten up just like any other organic waste. (We will get to the food part, I swear.) You can also put some in the yard. In the garden. In the compost heap....Get the idea? I've not reached the stage of "too much" but I do know in some areas of the country ash is not recommended for yards or gardens. But I also know people in these areas who have been doing it for years and *swear* by it. I think this may be another case of large scale, industrial agriculture studies where over-the-top amounts of certain organic materials were placed in a stripped field and then "run off" was measured and it was determined that these large amounts of organic matter were environmentally harmful and the next thing you know, regular gardeners are told not to use it. There is enough anecdotal evidence to support adding moderate amounts of wood ash to your garden and yard with no ill effects. By all means discuss it with your local extension agent or certified master gardener, but you also should not expect them to know everything or always have an open mind. They are just regular people and some might have their own biases. So make sure you ask some experienced gardeners in your area for their take on it, too.

This is what I do know: I've been using ash on anything and everything that grows green for about forever with no harmful effects on plants, trees, grass or me. I'm hard pressed to think of anything organic besides oils and salts that I wouldn't use to at least add to the regular compost pile if I could get it in small enough bits so it breaks down fast enough (oh yeah, clay and sand-like cat litter excluded, too) Oh, those black bits of "unburned" wood? They ROCK. Like little slow-release carbon pellets. Your roses will love ya even more. If you still have doubts but can't put your finger on just why that is, then just use a little and compost the rest.

Just because you may not be able to dispose of all the organic waste you personally generate in your one little worm bin, don't hesitate to find another creative solution for it. The main thing is to keep your organic waste out of the landfill. I'm begging you.

Soil and Compost

Have you ever done something one way your whole life and then after twenty years or so, you read some authority that tells you that you need to do it this other way to get a good result, but you've never done it that way and are getting a good result so now you're all confused? You know the "placebo effect" they use in medical studies? Where you give someone a fake pill but they still feel better even though it's not a real pill? I read awhile back that someone finally studied where that idea came from and found out it came from one flawed study done a way long time ago and everyone else after just kept quoting the study and then quoting the papers that quoted the study and the next thing you know, it was "common knowledge".

Well, that is how I think the whole idea of adding soil to your compost or vermicompost came about. I think the original idea was to add some microorganisms from the soil to the pile to "get things started". That may work, I don't even know how to find out if it really helps (as in: makes a real difference) or, if so, how much or how little actually works. Everything I've ever read is only as specific as "make sure you add a handful of soil" when you start your pile or bin. You can even spend some of your hard earned money on "compost starter", sold with the promise of containing billions of beneficial bacteria...Inside an air tight plastic bag. Hmmm.

I've never done this for any kind of composting endeavor and never had any problems that didn't involve my own laziness. The microorganisms that help break down the food and bedding in your vermicompost bin are naturally present in the food and bedding themselves. So if you want to add a handful of soil, you won't hurt a thing, but I'm not going to tell you that you should, because I've yet to find a reason to other than for grit (which I'll cover in the section about feeding).

What you don't want to do is dig a bunch of dirt from your yard and then put it all in a bin with worms and say you're good to go. Go back to the part about putting your worms in the garden or yard and re-read that. They need food. Composting worms may pass some soil through their bodies, but it is not "food". Some types of worms do eat dirt, but the ones you will have in your bin will want decomposing matter to eat, not plain dirt.

Compost, however, is a different story. Compost is not soil. Compost is decomposed organic matter. The nature of compost makes it one of those few things that can work for both food and bedding. The "bedding" part depends largely on just how "finished" your compost really is. If you are a lazy cold composter (cold composting being essentially what happens when you fail to turn and otherwise properly maintain your compost pile), then chances are good that if you put a handful or more of that in your bin before it is actually finished, it will heat up a bit once you start watering and aerating regularly to maintain your worms. Bad idea. Properly finished compost can be used of course, though I'm not sure why you would bother. The same stuff you use in your worm bin for food and bedding is the same stuff you would have originally put in your compost bin; not counting green grass and the like. So why not just put it in there to begin with? Cut out the middle man, so to speak. You will still be recycling your organic waste, with much less effort and a much greater return. This falls into the category of "no brainer" in my world. Yes, there are exceptions to this if you are pre-composting fresh manures or if your household generates more organic waste than your current worm population can handle. But for most people, the worm bin can handle their household organic waste without any pretreatment.

Paper, Paperboard and Cardboard

Ahhh, we've finally reached worm bin bedding nirvana! I told you I'd save the best bedding option for last and here we are. At least the best for an indoor bin. Look, I totally understand the excitement that comes with getting your first worms and especially the excitement from getting your first *harvested* bin of poop. I mean really, look what happened to me. I just can't stop. There is, however, a method to the madness of how this is all presented. Believe me, if I didn't include everything you would be asking me about it later. Worm composting just makes your mind work that way.

I know I've talked about a lot of different types of bedding. Try to remember that you want to *start* small. Get your feet wet with a smaller bin so you can get the hang of it and then move on to the bigger ideas, okay? A worm bin and paper products go together like water and wet. Don't start with anything else, really. It fits all the criteria for the worms to be able to thrive plus my special two extra criteria: free and readily available.

If you don't regularly recycle paper and such, take just one week to save all the paper, paperboard and cardboard you would normally throw away. Just one week. Put all your post-it notes, paper towels, empty cereal boxes, price tags, soda cases, mis-used printer paper, junk mail, magazines, and newspapers. (And anything else you can think of besides toilet paper---Yes, I know I say "poop" a lot and laugh when people cringe at the word, but even I draw the line at recycling your toilet paper. Maybe some day later. Right now I'm more worried about the seven pages it takes to print up that forwarded email.) Go ahead, do it for just one week. I'll wait.

Okay, did you do it for a whole seven days or were you getting a little overwhelmed after the third day or so? There are other books and internet articles and the like about how to use dishcloths and cloth napkins and such to help cut down on a lot of this waste, so I won't go on and on (this time), but I hope you got the point. Yes, a lot of this can be recycled at your local recycling center or even picked up and hauled away by your friendly neighborhood trash service. But think of all the extra effort and gas used for that. What if you could just recycle all of that right at your own house? Even better: what if you could recycle all of that right at your own house and *get something super cool for your efforts?* That, my friend, is exactly what you are about to do.

Not all paper products are created the same. Some are better for vermicomposting than others, though really any will work except for heavily wax-coated cardboard (which you don't run into very often outside the produce department). I get a LOT of questions about paper types in my classes, so we'll cover this pretty thoroughly since you may not know yet that you are going to be asking later if I don't. Let's start with paper and then move onto cardboard.

The most common papers you have in your house to recycle are newspapers. I'm not aware of any newspapers that are not printed with soy ink these days, even the colored inks. This makes newspapers a great bedding material.

The question of whether you can use the glossy paper advertisements that are in with most newspapers is one of your preference. My worms eat this paper up just like the rest of the newspaper. The preference part comes into play in the event that you are attempting to meet the legal standards for organically raised food. If you are then, at least as of this writing, you are not allowed to use the glossy papers in your compost until they update their rules to fit modern paper-making methods. Yes, it is still "organic" if you use these papers, you just won't be able to legally sell it as "certified organic". Composting or worm composting these papers is, more importantly, both *sustainable and responsible* so please bear these equally important words in mind when deciding what to do with those papers.

Junk mail and computer paper are the next most common papers you will have access to. Junk mail can pose special problems because so many of the mailings have something plastic inside or they have a plastic window on the envelope. The plastic stuff and the envelope window cannot be put in the worm bin. Well, they *can* be put in there since they won't hurt the worms, but believe me, you will be very frustrated trying to pull all these little plastic pieces out of your finished compost. It's a pain in the rear, so save yourself the trouble and make sure you remove these things from your junk mail before you use the rest of it for bedding. Or better yet, a simple internet search will give you the addresses and phone numbers you need to automatically opt-out of getting the vast majority of this junk in the first place.

Computer paper is totally also fine to use as bedding. Some people don't like to use it because it is bleached paper. The worms don't care about this. The amount of bleach left in the paper is negligible. Not to mention that there is chlorine in your city water and in more of the bottled waters than you may want to believe. It's fine to use, don't worry about it unless you truly want to. You do have other recycling alternatives if this is a personal sticky point for you. Toners for your printer are largely made from carbon and/or soy based ingredients, so there is also not a major concern with these, either.

Regardless of the type of paper you use, you will need to turn it into bedding by shredding it. Lazy girl here totally recommends just putting it all through the shredder (which is going to turn those plastic bits in the junk mail into even smaller and exponentially more irritating bits so really, make sure they are taken out before you shred). The paper shredder is one of my most favorite inventions ever (after the automatic dishwasher, of course). What a time saver. And in this era rife with identity thieves, how much more safe can your information be if it goes through the shredder and then through the worms? Talk about security!

I have a shredder right in my kitchen and the junk papers can go right in there as soon as I'm done with them. I also, because of the scale I raise worms on, have bags (and bags and bags) of paper from other people. Be careful if you take shredded paper from other people as not everyone will be as diligent as removing the plastic bits as you will be. This is the trade off you must accept when you decide to use someone else's shredded paper. I haven't found a way around it, no matter how much I ask people to please not put in anything with plastic on it, it still happens. Just keep that in mind when you decide you need more paper and don't have enough of your own. You may just want to ask for newspapers or such and shred them yourself to eliminate this problem. More work for you, but less having to pick out plastic bits.

You can absolutely shred your newspapers and computer paper by hand. Narrower strips tend to be better in the long run, but you can get away with hand shredding in just about any configuration that makes you happy. Some people even find shredding paper for a half hour a week sort of meditative. It doesn't require any brain work whatsoever and gives them some "quiet time" that is also productive. If you want to shred by hand, have at it. Just make sure you are not wearing light colors when you are shredding newspaper as the ink will get all over your hands and eventually on your clothes.

Paper towels are probably a big part of your paper waste stream if you have little kids. You can totally use paper towels in your bin for bedding, but the problem is putting them in there whole. They tend to get all bunched up and end up being these super hard little paper balls that the worms can't really get chomped down very well and sometimes not at all. If you can figure a way that works for you to shred your paper towels, then by all means do so. For most people this is too much of a pain. I totally believe we all need to reduce waste on a much larger personal scale than most people are currently doing. I also recognize that you may actually have a life and shredding paper towels is likely not a priority. If that's you but you also want to reduce your waste stream then please explore some of the easy alternatives to paper towels like cloth napkins and reusable rags. I totally understand that it is sooo very much easier to just "throw it away"; but **please** remember: *there is no "away"*. It may take a little getting used to, but generating excessive waste is not something we should want to be "used to" either, right?

Paperboard is the step kid of paper and cardboard, but not the evil step kid. It's not quite paper and not sturdy enough to be called cardboard. Paperboard is the stuff your cereal box is made of, your tissue box, your pasta box, the toilet paper roll...and the backing of all those other things you buy that have that molded plastic front glued over the product you are actually trying to buy. By and large you will treat paperboard just like regular paper. You can run it through the shredder just fine, too, but usually just one "sheet" at a time so you don't burn your shredder out. Paperboard is different from paper in that frequently (mostly for the products you buy with the molded plastic front) it has some eye-catching metallic design that sometimes should not be used. I have bought a certain brand of mascara that has three strips of bright silver across the paperboard backing it is packaged in. Two of these strips are actually overlaid onto the paperboard with something like mylar and won't decompose. The third strip is "painted" on and decomposes fine. When I get things like this, I put it in the regular recycle and let the big time recycling guys take care of it. Their process is much more sophisticated and they can work with it better than I can. Besides, I hate picking out bits of inorganic crap from my worm, um, crap.

Cardboard is awesome for worms. Corrugated cardboard is actually made from three layers of cardboard (two flat, one wavy) that are glued together. The glue is like worm crack and they love it. The little corrugated areas are like little worm love shacks and they really love that! (We will get to that a little more when we talk about the birds and bees of worm life.) The problem with it is that most of the time corrugated cardboard is too thick to go through a regular shredder and too tough to go through most shredders often without quickly burning out the motor.

Instead I like to use a whole piece of moistened cardboard on the top of my bin to help keep moisture steady, along with a piece of plastic, which I'll talk more about in a sec. The worms will only eat on the parts of the cardboard that are moist, but they eat and eat at the glue first and you will quickly find that the cardboard is starting to separate. Then in a week or so more it will start to get eaten in the middle. Then the next thing you know you need more cardboard!

A friend of mine who just has a single bin and more waste than her worms can eat, likes to save up her cardboard and egg cartons and such and, while she is watching her favorite show every week, she tears it all up in wee bits and brings what she can't use to me for my worms. What a perfect set up! She doesn't knit, so tearing up the cardboard gives her something to do but still lets her concentrate on who is getting voted off the island and why. We all have to have priorities.

The great thing about all of these kinds of bedding is that you can totally use a mixture of any or all of the things on the "acceptable" list. Please start with just paper and cardboard and finish the book before you decide to add any of the other bedding types to an indoor bin. Sometimes non-paper beddings bring non-worm guests. By "non-worm" what I mean is "unwanted". So do please finish the whole book before you make any decisions you may later regret.

Put a Lid on It :

Covers for Your Bin

You have your bin and you have your bedding, but most of you will also need some kind of lid. Again, if you have house cats or litter-trained ferrets or any kind of dog breed that rhymes with "Rabrador letreiver", you *will* need a lid for your bin if you plan on putting it any place these little rascals might have access to it. My cats can open cupboards and drawers (the wily little beasts), so these are not animal-proof if your pets have brains. (My lab, God love him, was apparently born with only a part of a brain and doesn't know how to get into a cupboard so if you have one of his siblings, you are safe.)

If you followed my instructions and have a plastic bin to start with, it likely came with a lid unless you got so excited about raising worms you left the store without it. No lid? No problem. We are very good at improvising.

I'll get more into the whole "holes in the bin and lid" issue even more when we get to the bug section. (Hey, quit freaking out about the bug section already. It's really not as bad as you think.) For now, I'll only talk about the lid itself, not the bin.

First of all, unless you are building a box to put outside where the birds can get to it and as long as your bin is not available to the aforementioned indoor animals, you don't actually "need" a lid. I do, however, think a cover of some kind is a good idea regardless.

Let's start with those of you who remembered to grab the lid that goes with your plastic bin. Yes, you can go get your power tools now. You're welcome.

Most folks will tell you to just drill a few holes in the lid and be done with it. I personally like to make things a little more secure for the long haul. I've found that the "where" of your bin location will likely change over time. You move, you get a roommate, a spouse, little people, whatever. Life sometimes gets in the way of your personal long term plans. As such, you can wish later that you made a different lid or you can fashion it right in the first place and it will always be a good lid and never a burden or a mistake you can't change.

My favorite thing to do with a plastic lid is to get out my drill and my reciprocating saw. Don't have one? Ask around. More people than you think would have these items and are more than happy to have an excuse to get them out. The vast majority of plastic bin lids have a natural depression in them two or three inches from the edge. This depression totally rocks because it serves as a natural guideline to whoever is wielding the power tools. Take a large drill bit (sorry, no specifics here, large = the largest drill bit you own) and drill a hole in one corner of this depression. Now drill a similar hole in the corner opposite the hole you drilled first. Now go put the drill away.

Next take a reciprocating saw (the smaller toothed blades tend to work best) and, with the lid *firmly anchored* (this works best if you put it on the bin itself before you add the bedding), cut out the majority of the center of the lid, starting at the holes you made and then following the line of the depression. It doesn't have to be perfect. You are not entering the lid in an art show. If you do this right, you will have a rectangle of plastic cut out of your lid. Don't throw the piece you cut out away. Your worms may need it. You can use a hacksaw for this step if you like, but you still have to have somewhere to start your cutting so a drill is pretty much a "must".

Now, find yourself some window screen. If you have mice in the area where you are going to put the bin, make sure you have metal window screen and not fiberglass. Mice can chew through fiberglass quicker than you can say "supercalifragilisticexpialidocious". And they will, but I'll save that for later. Get out your staple gun. Ohhh, more tools. Arrr! Arrr! Arrr! Put the lid on some concrete or other stable surface that is NOT your linoleum kitchen floor (not that I know this from experience, but the staples will go right into the linoleum). Now staple the window screen onto the *underside* of the lid. This matters. If you do it on the top, the shape of the lid won't allow you to get a good staple in. Plus, it will look trashy. Yes, yes, I know: not an art show. But still.

I like to cut the screen a little bit bigger than I need and fold the edges over at least twice and then staple. This really seems to create a better barrier. Don't be shy with the staples. Now, turn the lid over. You should essentially have a window in your lid that is fully covered by the screen. Gorgeous, isn't it? Such a set up will help with both air flow and will allow excess moisture to escape your bin via evaporation. But it will also keep out the cat and flying bugs. Voila. You are a genius. Don't have any screen? If you have some landscape fabric or some very loose weave fabric, you can still make a very nice worm lid using the exact same techniques as with the window screen.

Because my inside worm bins are in a breezeway that is also accessible to flying bugs, I also like to put a bit of self-adhesive weather stripping around the top lip of the bin itself. You can put it there or on the underside lip of the lid. I just like to put it on the bin itself because this way I know it will create a good seal where I need it. Clean the bin edge with a little rubbing alcohol before you put the weather stripping on to get any manufacturing residue off the bin, first. Now, I realize everything I just said may make you think that I am contradicting the whole idea of not compromising the bin in case you want to use it for something else someday. Not at all. It is correct that I don't want you to drill holes in the bin itself.

Those holes are very difficult to plug up properly if you do decide that you want your worms somewhere else and want that bin back to store your winter clothes in instead. But cutting out the middle of the lid and replacing it with some screen will not hurt the future purpose of the bin for the vast majority of the things you would want to use it for in its next life. You can totally keep moths out of your woolens and the cat out of your wrapping paper and ribbon with a screened lid.

For those of you loathe to "hurt" any part of your bin, consider this: one of my students improvised her own lid in the most brilliant way with a little bit of landscape fabric. She took my "you may need the plastic bin for something else in the future" warning to heart. She decided to cut only three edges out of the middle of the lid. This left the lid with a "hinge" on the uncut side, basically creating a flap in the lid. She stapled landscape fabric to the underside of the lid, a la how I described stapling the screen to the lid just a second ago. She then put a half-inch wooden dowel under the "hinge" on the uncut side to hold up the flap, which propped open the flap enough to allow sufficient air for the worms but also still preserved the integrity of the lid. Ta da! A great lid for her bin, but it also keeps the lid intact in case she decided to change the purpose of the bin from worm home to something like "holder of sweaters in the off season".

If you have no lid at all for your container, you can still improvise a lid. An improvised lid, however, is not the best bet to keep out indoor beasts, so if that is your main concern then it might be a better idea to purchase a new bin for your worms. Just don't walk out of the store without the lid this time.

The least expensive type of non-lid cover is simply a piece of cardboard. Corrugated cardboard is best, though you can use two or three layers of paperboard if you like instead. Soak the cardboard in a little water for just a few seconds. The cardboard shouldn't cover the entire surface of the bin, leave a few inches around the sides or just put the cardboard over half of the surface. This will allow for some evaporation in case your bin is too wet; a subject I'll get to in much more depth very soon.

Now find a piece of thick, rigid plastic that is another two or three inches smaller all around than your piece of cardboard. This will go on top of the cardboard and keep that spot moist for those times when your bin is on the dry side. This way there will be ample opportunity for the worms to be happy with their environment because there will always be an area that will be "just right" for them. I like to use a bucket lid for the piece of plastic. Kitty litter sometimes comes in buckets. Those lids are great. A lid from just a regular paint bucket is totally. Some of those gusseted, "they're only a dollar-let's get two!" reusable grocery sacks have a removable plastic piece at the bottom. This will work, too. If you did have a lid to begin with and cut out the center, this center piece will also work just fine, though likely you will have to cut about a third off of it to keep it at a reasonable size.

After about a week, it is very likely that every time you lift up the plastic piece to feed the worms, you will find that there are a zillion worms under the plastic and that your cardboard is gone wherever it remained damp. The worms LOVE the spot under the plastic piece; they will still roam around to the other parts of the bin, too, so don't worry that they won't eat the food you give them. But also don't think that means it's a good idea to cover the whole thing with the plastic because trust me, as much as they like the plastic they will still need air. If you lift up the plastic and find that there are dying and dead worms under the plastic, it usually means your bin is way too wet or totally overfed and we'll cover how to correct these things quite soon. Because the worms like the plastic and cardboard so very much, I put them both in my bin even if I do have a proper lid; which is exactly why I had you save the center of the lid when you first cut it out.

If you are using a bin without a lid, it is usually better to also put some screening over the top of it. This matters more in the warmer months when annoying things that also fly can be a problem. You can still cover your bin with screening even if you don't have a lid to staple it to. Just cut a piece of fiberglass screen (aluminum screen won't work as well for this kind of cover) about four to six inches longer on all sides than the size of the top of your bin. Secure it to the top of the bin with clothespins or binder clips, making sure to double up on the corners. You will want it to be so much longer than you need so you can also fold up the sides a little as a bit of an extra deterrent in the warmer months. You can also put adhesive weather stripping along the top of the bin if you find this is really necessary. That is about as good a seal as you will be likely to get without a real lid. But if you add up all you might have to purchase to make a lidless bin work well for your situation, it nearly always comes out less expensive to just go buy a new eight dollar bin.

I already talked about how to cover a trench or windrow when we talked about how to make them. No need to "cover" that again (haha). Next we'll learn about how to maintain the whole shebang. Yes, yes, I know you are chomping at the bit and wanting me to talk more about the worms themselves. I will, I totally promise I will. But if I started there instead, you would already have your worms and would have put the book down so you could "keep an eye" on them and the next thing you know you have screwed the whole thing up and will have wished I'd written more at the beginning about how to set up and maintain a bin. Not, of course, that I know this from personal experience. There is a method to this madness, I promise. By the end of this book you will be filled with that quiet confidence you can only have when you know exactly how things will turn out.

Cool Breezes and Worm Rain:

Living Conditions

Okay, you've got your bin and you've got your bedding, now what? There are two more steps to take to make the conditions good before you add your worms. The first is making sure the moisture and air levels are appropriate and the second is to get the food ready. Food is a huge category and we'll cover that all on its own.

I'm going to get pretty detailed about the moisture, but it all only comes down to this: make whatever type of bedding you have as wet as a mostly well wrung out sponge...And keep it that way. That's it. You can just skip to the food section now if you want. But make sure you mark where you left off since, for whatever reason, that simple "sponge rule" is never quite enough. I'm pretty sure the problem is the "and keep it that way" part.

The reason it is hard to keep the bedding at the proper level of dampness is because of both evaporation and gravity. Most of what you can do to stop evaporation would make your worms every kind of dead because it would also cut off too much oxygen. No oxygen, no worms. Gravity is great for what I hope are obvious reasons ('cause for one it will keep your worm poop in the bin!). But gravity also sucks (ha!) because it will pull the water from the top and middle of your bin all the way to the bottom, where, if you don't correct for this problem regularly, it will displace all the oxygen there which in turn will cause anaerobic conditions and as such, some serious stinkiness. But let's start at the beginning and work our way to how to keep it damp-sponge-like for life.

Whatever you are using for bedding needs to be properly moist *before* you add worms. Moisture can come from a number of sources, but plain old water is the best source. You can use old coffee or tea if you like, but try to limit using fruit juices and beer and wine or sour milk as a major component of your moisture for your bin. Anything besides plain water is really more like food for your worms and you *can* give them too much. Using plain water leaves them areas of bedding where they can get away from the food bits if they want or need to.

Plain water, unfortunately, is not always as easy to come by as you might think. Some municipalities use a large amount of chlorine to treat the water that you get out of your tap. If you have highly chlorinated water, make sure you let it sit out for a couple three days and give it a stir every now and again to aerate the water, or give it a good boiling and let it cool to help some of that chlorine go away. It isn't so good for you, either, but it can be very bad for your worms.

If you have hard water, that won't hurt the worms, but if you have hard water and use any kind of water "softening" system that relies on salt, then you will need to find another source of water for your worms. Excessive salt is incredibly bad for your worms. Salted water falls into this category. Good thing for you that you won't need much water for your worms. "Borrow" a couple or three jugs of water from a friend who doesn't have these water issues and use that water for your worms. Heck, borrow some water for your ownself while you're at it.

I also get questions about using soapy water. Soapy as in leftover dishwater and the like. This is one of those times when I must say: I don't know. Mostly. I do throw my dirty dishwater into an old tractor tire I use as a bin. Worms thrive in there and it is an easy way to get water to it since the hose is nowhere nearby. But I've never used this same water in an indoor bin. I'm not worried that the soap will hurt the worms, but am concerned that it might hurt the microbes. There are a lot of odd chemicals in some soaps. Especially the "anti-bacterial" kinds of soaps. The outdoor worms I give my dirty dishwater to have lots of places to go to find a different area if they object to the soap. Worms in a normal, indoor bin, do not. I'll let you decide if it is worth the risk for you to use this until more research can be done.

When you get the bedding ready for a new bin, ideally it should be prepared a whole week or so in advance of getting your worms so it can be slightly pre-decomposed. For things like paper products (*most* recommended for your first bin and for indoor bins in general) and straw and leaves and coir and the like, putting them in a bucket or large plastic bag and then wetting everything down well works great. I also like to add a cupful of coffee grounds or used tea leaves and some grit to the bedding and mix it all up with the bedding while I'm at it. We'll talk more about grit in the food section. Coffee grounds and tea leaves are small and decompose pretty quickly after they are used and it provides a little bit of ready food for the worms right from the get go. When you wet everything down in the bucket or bag, don't try to fill the bucket with water, just fill it with the dry bedding and add a bit of water, gently mix it up, add a bit more bedding and a bit more water, cover the bucket or close off the bag and then go read a book for an hour (I recommend *this* book!).

When deciding what kind of bedding to use in your indoor bin, keep in mind that I do *not* recommend using livestock manures in an indoor bin, though you certainly can. (But not fowl poop.) Part of why not is that they can have a bit of an odor at the start. (Yes, yes, I do certainly know that for certain people the smell of livestock manure is like high end French perfume. But even those folks might not like that smell in the house all the time.) The other part has to do with certain bugs, which we'll talk more about in that section. If you are dead set on using manures in an indoor bin, and frankly I cannot imagine why, start with just a little mixed with paper or the like and add more as your worms progress through what there is to eat. Vermicompost can really help mitigate bad smells, but wet animal poop doesn't always cooperate with that until the bin is at least about a third converted to worm poop.

After an hour or two of just letting the bedding soak up the water, or even the next day (better), the bedding will be all nice and moist. Most likely, especially if you leave it overnight, the stuff in the top few inches will be *almost* wet enough but not quite, the stuff in the middle will be *just right* and the stuff on the bottom will be way too wet and will be dripping with water; kind of the Goldilocks version of worm bedding. Put all the top two thirds (the almost-right and the just-right stuff) into your bin. Pick up the super wet stuff from the bottom and wring it out semi-well. On your final wringing, only 5 or 6 drops or so of water should come out.

When you wring it, it will get all mushed up if it is paper or the like. No matter what type of bedding it is, put it all in the bin and then fluff all of it up together, breaking up any clumps of damp bedding. You really don't want any big clumps, but you won't get all the small clumps out the first time around so don't sweat this too much. After you mix it up, you should feel that the bedding is both fluffy and about as moist as a wrung out sponge and by the next day only 2 or 3 drops of water should come out if you squeeze some. That "2 or 3 drops" is your ball park figure for the life of your bin so get used to how it feels because once you add your worms, you will not be doing any squeezing unless you want to smoosh any worms that are stuck in the bedding you chose to wring. Please don't over think the moisture and work and work to get it exact, just get in the ball park. A little bit more wet is not a big hairy deal. A little more dry could pose problems if you are not diligent. If you are going to error, definitely error on the side of slightly more wet. And by "slightly" what I really mean is: *slightly.* Don't get crazy with the "slightly" and think "wet" is only "slightly" more saturated than damp. It isn't.

Sometimes you will find internet worm people who think it is necessary to purchase a moisture meter for your worm bin. This item is generally sold in garden centers and has a metal probe with a sensor that you stick into your houseplants to gauge how wet or dry they are. They usually cost between seven and fifteen bucks. I've said it before and I'll say it again and again: it doesn't make a ton of sense to purchase a bunch of stuff to help you recycle. If you already have a moisture meter for your plants and like using it, then by all means feel free. I find that my fingers work just fine and I can just stick my finger in the bin to gauge for my ownself how wet or dry the bin is. But I also do this for my houseplants so maybe it's just me. If you don't have fingers, though, then a moisture meter might be a good investment.

You will want to fill your bin with bedding to between 8 and 10 inches deep but only for the very first time you make it. Try to always keep the top of the bedding at least two inches below the top edge of your bin. Things will settle down in a couple of days to be between 6 and 8 inches deep and that is the level you want things to be at for the life of that bin. (If your bin is only 6-8 inches high, then fill it to within an inch of the top. If your bin is shorter than that, you may want to get a taller bin.) If the bedding is too high, then when you are doing your feeding and maintenance things will spill over the side and make a mess. Much less than about five inches deep of paper bedding and your worms start getting unhappy and you have to spend a lot more time on maintaining the moisture levels. It can get that low with few problems once it is at least half converted to poop, because the poop will hold a lot of moisture itself. But until that time you don't want it to get so low unless you really want to be messing with your worms every other day. Again, don't over think this, just try to get there more or less.

You will want to keep some dry bedding near your bin. The vast majority of people start out keeping things far too moist instead of just a little bit extra damp and having the dry bedding nearby is a very quick fix for when you find you got things too wet. For a single, normal sized bin, just keep a small bag or empty coffee can full of dry paper shreds near your bin and that way you will always have it when you need it and you won't fall into the "I'll get it later" trap, which too often leads to dead worms. It only takes a second and it's easy to do if you have what you need nearby. It's also a great idea to keep some "just moist enough" bedding in a coffee can or some such other covered container nearby for when you need to add bedding but conditions are pretty near perfect. The total mass you originally started with will be reduced as the worms eat it and you will need to add some more bedding periodically to keep the depth stable. We will discuss when to add more bedding more during the maintenance section.

Very few people will actually end up having problems with things being too dry. Since gravity and evaporation are constantly working to dry out the top part of the bin, though, sometimes is *seems* as though you are dangerously close to things being too dry when in reality things are only too dry on the very top. I'll talk about how to maintain your bin so that this is not much of an issue, but for the first couple of months into your worm endeavors, it is unfortunately *not* going to be amazingly rare for you to fall into this trap. Never add water or dry bedding unless at a minimum you stick your whole finger into the bin and *all the way* down to the bottom. No, it's not gross, but you will get a little dirty so perhaps when you are in your tuxedo about to leave for the symphony is not the ideal time to check.

If you feel adequate or excessive moisture in the middle and at the bottom, then do not add more water. If things are really wet all over and you can smell some stinkiness starting but don't have time to do the proper maintenance, then at least put some dry bedding on top and/or take off the lid and leave it off for a day or so if you are able to (because you don't have house cats or the like with access to the bin) until you can take the five or ten minutes you'll need to do proper maintenance.

If you have some kind of lid on your bin and you are not adding much water but your bin is still regularly too moist, then it may be that you have "too much lid" and not enough air is getting in. If you have worms trying to escape or crawling up the sides or constantly up on the lid and in the grooves of the lid, or actually escaping and dying on your floor, then your worms are not happy with something. Composting worms don't try to leave because they are naturally mischievous or rebellious, they only try to leave because conditions are bad and they don't want to die. Which is exactly what *will* happen if you don't correct the problem for them. By this I mean: right away! They don't know that if they leave the bin they will not be going to a nicer part of the "yard" and instead will wind up on your floor, where they will shortly turn into many different configurations of worm sticks, they just know that where they are sucks bad and they gotta escape and find something better. Usually it's the food, which we are going to cover quite a bit in the next section, but the next possibility is that they do not have enough air so you will want to fix that so they can get the right amount.

As I mentioned before, if you looked for some instructions on the internet on how to build a bin, they likely told you to put all kinds of holes in the bin and the lid to allow for air to get in. Lots of those directions never specify how big the holes are either, which I find quite hilarious since "drill some holes in the bin" could hardly be much less specific than just "round". I've had people bring these bins to my classes with all sorts of holes in the bin. Some decide to put about a hundred or so holes about the size of a pinhead all over the bin. I can relate: I get a little carried away with the power tools sometimes myself. Some make quarter inch holes in the lid and in the very bottom and all around the top half. Some can't seem to decide so they put little holes in the bottom and big holes in the top. Or vice versa. Here's the solution to that confusion: *don't drill any holes in the bin at all.* That would mean: ever. I know, I know, you already did. My condolences. You may want to just start over. That decision will depend on how much of a life you want to have outside of your worm bin. Here is the problem with the holes, regardless of the size or placement:

I already mentioned that one problem is if you put the holes in and decide to go to another type of bin, now you have a bin with very limited use because now there are all these scratchy holes in your previously-nice plastic bin.

The bigger problem is when you drill the holes in the bottom or lower sides. Yes, your worms can and will crawl out of those holes. Even the little holes. I may have mentioned that baby worms are not so bright. They are also tiny. But that is not the biggest part of the problem. The reason the faceless and often nameless (coincidence that it rhymes with blameless?), internet people tell you to drill holes in the bottom of the bin is because the number one problem people experience the first few weeks they have their bin is too much moisture. The holes in the bottom allow the moisture to escape. Not eensy droplets either. No, that would be too easy. Depending on how badly you erred in adding water, it could be anywhere from a few drops to a few *cups* of water. Not clean and clear tap water. Water that ran through both worm poop and rotting food. Yep, it is about as nice as it sounds.

Worm poop itself actually smells wonderful. Exactly like the best soil you ever smelled. Like spring, like clean, a little even like sunshine itself. One commercial grower describes it as "like walking in a forest after a rain"; is that not the most gorgeous idea?

But there are bacteria living in the poops. *Good* bacteria for sure, but bacteria that will get every kind of icky when you remove the air they need to live. Then, the evil twin bacteria take over. "Too much water" will weigh things down and will actually *displace* the oxygen. Remember: gravity is our friend, but not when this happens. This is what makes the bad smells I keep talking about. When your bin is working the way it is supposed to, you will be dreaming of seed catalogs even if you have never planted anything in your life. It smells that good and is that powerful. But when it smells bad, you will be cursing your worms and wishing you'd never even *heard* of this book. Well, I certainly don't want that! (Umm, the book regret part, not the worm cursing part...Your worms can't actually hear you and even if they did, for all I know your cursing could be sweet nothings in worm talk so curse away. Just don't wish you never bought the book.)

Water that went through worm poop and rotting food and picked up those bacteria and then came out of those little drilled holes and sits in whatever you put under the bin for the express purpose of catching this water will begin to get a bit of an odor after about 10 or 12 hours. This may be the original origin of the word stagnant (it just *sounds* bad, doesn't it?). Depending on how deep the water is (hope the same website that told you to drill holes in the bottom of the bin also had you put something relatively deep underneath it...if they didn't, that may be another good reason to stay away from the nameless), the smell will be either readily wicked or, if there isn't too much, it will be one of those insidious yet heinous smells that you can't quite figure out the source of but that will still drive you completely insane. Yes, I realize if it is dripping into a tray it will be exposed to air. But only on the top. Unless you build in a motorized device to stir the liquid that leaks out, it will go stagnant and it will stink. Period. Frankly, there is no way to make it nicer except to eliminate the holes in the bottom. Or make sure the bin doesn't have so much water that it leaks out. But if you're careful with the water, then why do you need the holes? See what I mean? No holes in the bottom, okay? But what about the holes in the sides or the holes in the lid? I'm going to go over holes in the lid and sides in the bug section, so for those you have to wait, sorry. All will be revealed in due time, I promise.

Now before I went off about the holes in the bin, I was talking about your worms leaving because conditions were bad. They usually will not try to leave if things are too wet. It's a food or an air problem if they are leaving, nine times out of ten. The one in ten is it's going to be a stinky water problem. Your worms will most often will hang out in that stinky mess until, before they know it, they are overcome with the lack of oxygen and die right there, contributing to the stink as they decay. If you didn't notice the stink before, you will once the worms die. Don't worry, it won't happen in a day or two and you will be in your bin feeding and doing maintenance once a week so the theory is that you won't be having to deal with this (especially if you pay attention to the rest of the book; so pay attention already). Do, however, keep this in mind if you are going on vacation for more than a week and think you will "just put in some extra water and food" and they will be just fine. I'm pretty sure just because you will be on vacation doesn't mean that gravity will be.

If your worms are trying to leave and it isn't a problem caused by disorientation from being shipped in the mail, the first culprit to go after is the food. Check to see that you didn't either throw in too much food or throw in a combination or amount that might be turning hot. A bunch of rotten spinach will act just like rotting grass. Too much grain will also get hot, especially if it is combined with the right amount of something sugary, like fruit. If you aren't sure where to start to keep your worms from crawling out and you can't tell if it is the food or not, then error on the side of caution and pull out at least half of the food (making sure there are no worms in what you pull out) and then mix the remaining food with some fresh paper scraps. Scoot the escapees back into the bin, put them somewhere that the bin is exposed to light and wait an hour or so to see if that does the trick. This, of course, is assuming that you did not use something like grass clippings for bedding. If there is anything in your bin other than paper shreds and the like, go back and read that section again to make sure the bedding itself is not the problem.

If fixing the food and/or bedding doesn't halt the exodus, the most likely suspect is going to be air. Not the fact that there is air, obviously you can't change that! No, I mean that there is not enough fresh air. This can be caused when the bedding is just so wet that it gets all smooshed down and compacted, but generally the lack of fresh air is because you have too much lid. You can easily check this by simply removing the lid and seeing if that does the trick. Fluff up the bedding while you are at it just to make sure it is not actually all smashed down. If the lid is the problem then have no worries! There is a whole chapter on that written just for you. Feel free to come right back here after you review that bad boy one more time.

Worms will also not usually leave an indoor bin because things are too dry. They need to have moist skin to breathe, but also to move very far. If you do end up being one of the rare ones who really does have a bin that is too dry, that is an easy fix. Just add water. Regularly, if this is a regular problem for you. I had this great couple in my class who kept their lidless bin in their bathroom. They didn't have house pets so having no lid was not an issue. They kept a squirt bottle of water in the bathroom and every day they gave the bin a few squirts on the "mist" setting and this, combined with the humidity in the bathroom and the natural moisture o the food, kept conditions great for their worms. They labeled the squirt bottle...drum roll please...Worm Rain. Is that not the funniest thing? I also had a student who made the mistake of having a super busy life. Okay, sorry, that is not the mistake. By ALL means, have both a life *and* composting worms. But she did not budget time for her worms, which is especially important in the first few weeks. They need at least 5 minutes a week. Sometimes a whole ten, but mostly only five. If you won't make time to give them five whole minutes a week, you may very well end up with what she got: worm sticks. Then you will have to start over, like she did. Don't give up even if you don't get it just right the first time, okay?

How well your bin will hold moisture depends on many factors: is it plastic, wood, open on top, closed on top? I could go on and on, but hopefully you get the picture. No, sorry, I can't say that this bin will lose an average of X% moisture per day at relative Y% overall humidity or even anything close to that. It is totally going to depend on *your bin* and on *your environment.* There are always exceptions to every rule, right? That is why you will do regular bin maintenance when you are feeding: so you can get to know *your* worms in *your* bin.

Okay, now you understand the 2-3 drops rule and then some. It's important that you start out that way and we will totally get to how to keep it that way in a bin, promise. But what, you ask, does one do if you totally disregarded my advice and didn't start with just a bin but decided on something much grander like a windrow or other outside worm home? Well, shame on you. I'm not writing this just to hear myself type. (Oh no; I'm turning into the written equivalent of my mother!) I will just pretend that you did follow my instructions to the letter and have become so successful at bin raising that you want to try something bigger. There, now we can both live with your decisions.

If you are building an outdoor bin (an actual enclosed structure, like Grandpas), the rules for type and preparation of bedding are the same as for an indoor bin. You have some leeway with being able to add some manures since it is outside, but not big quantities of anything green since the worms will still be enclosed and unable to escape just like in an indoor bin.

If instead you are going to use a trench or windrow or the like, especially if you are going to use manures as your bedding, you are certainly not going to soak all the bedding and then squeeze all the water out. And I'd never ask you to. But outdoors you have a lot more leeway than you do indoors. You also have gravity and evaporation problems much larger than the proportion by which your outdoor endeavor is larger than an indoor bin.

Most outdoor windrows (and though I will say windrow, the rules are about the same for a trench) will be made of livestock manures or yard waste or crop waste or some combination of all of the above. I do not recommend using paper shreds as the major component of your bedding in a windrow, since it is likely that some will dry out and blow all over tarnation and make a big mess. If you do, put them closer to the bottom and use something on top that is less likely to irritate the neighbors if it blows in their yard, like leaves maybe. Paper shreds in a trench are more likely to stay put, though in either case I still avoid putting them on the top layer, if I put them outside at all. But then again, I have super-cool neighbors so for me it's not really a problem.

I have not found it necessary to pre-mix your windrow ingredients. If you have, for instance, leaves and horse poop (remember to age or at least pre-compost your manures) and some old hay, just layer them as best you can. Don't break your back trying to get everything evenly mixed like you might for a hot compost pile. I do recommend, if you have some handy, putting a few sticks (larger pine cones will work, too) and such every other layer to help prop things up and keep it all aerated. If your windrow is only a foot wide and a foot deep, though, you can skip the sticks. I absolutely think building your windrow or filling your trench *before* getting any of this wet is the smart way. It makes the work far less back breaking when everything is dry first.

As you are building your windrow you have a couple of options for moisture you need to think about before you are done. I highly recommend investing in a couple of good soaker hoses per windrow, depending of course on the length of your windrow. You can hook up to six together without any ill effects if you have decent water pressure. I've found that with more than that, the end ones don't seem to have enough water. Just start embedding the hoses when the windrow is halfway built, then snake them back onto the top until you have used up all the hose.

I'm sure there is a technical name for the different kinds of rubber used to make soaker hoses, if only I knew where to find that information. Since I haven't, when you are shopping for either a new or used soaker hose (another great find on Craig's List), make sure to only buy either the fabric kind or the smooth rubber kind. For whatever reason, the bumpy rubber kind does not bend well and you will have a hose with far larger holes in it than you will want. Soaker hoses rock because they can be left outside all year and not be damaged from getting frozen. Make sure you put the connecting end at a place on your windrow where you will always be able to find it to hook up the yard hose.

Alternatively, you can set up a drip system or a mister, just like you would for your trees or garden. If you decide on a drip or mister, I do recommend that you do not make the windrow very high. Also make sure that your system will be able to be used when the weather gets cold if there is occasional freezing. Most will be fine if you set them up right.

Either of these alternatives are low maintenance. You may also decide to just water your windrow by hand, which you certainly can do. I do not, however, recommend this method unless you are retired. The problem with hand watering is that you will have covering on the windrows. Covering that will only get heavier when it is wet. Heavier and dirtier. A soaker or drip allows you to maintain your windrow without constantly having to uncover, water, and then re-cover everything. Plus, hand watering is far more time consuming. But if you decide this is the route for you, then remember when you are building your windrow to add water every 5 inches or so of depth in order to get the whole thing soaked well to begin with. When you are watering after your windrow or trench is well established, you will likely want to shove the hose in a few inches down every half foot or so to make sure the water goes down past the top of the bedding. Hand watering just the top doesn't always work well.

Some folks find it difficult to translate the 2-3 drops of water rule for bins into its equivalent outdoors. A windrow will be too big to do much regular mixing or maintenance with. Yes, you can turn it all with a shovel, but this is a pain in the neck (and back) and, especially if you have a soaker hose embedded in the windrow, can be overly time consuming. It is also completely unnecessary. Maintaining moisture levels outdoors is actually easier as far as I'm concerned, than maintaining moisture in an indoor bin.

I just put in the soaker hoses and turn them on for a couple of hours every 4-5 days in summer and every week to ten days in the winter, depending in all seasons on the rain or snowfall that week. (I should mention that in the part of Colorado where I live, we rarely have snow that remains on the ground for more than a week. If you live in an area that is covered in snow that never leaves in the winter, then you will certainly want to make sure your watering practices will reflect that. The hose should be buried deeply enough that it won't freeze and the whole shebang should be covered with enough straw or carpet or other insulation to make sure that not more than the top couple of inches will ever freeze solid.)

The soaker hose will generally only moisten the area where the hose is, out about 4-5 inches on either side. Your worms will hang out there and eat until that area is all nice and worm-poopy. Since worm poop holds moisture so well on its own, after a short time the amount of windrow watered on either side of the hose becomes greater as the water is able to wick through the poop and into the dryer areas, eventually making everything moist enough to be turned into worm poop. This whole process, of course, depends on you putting in enough soaker hoses for the size of windrow that you have. One hose for a two foot long by two foot high by twenty foot long windrow is not going to be enough. No, just like with watering an indoor bin I *cannot* tell you exactly how many hoses to use. That answer will depend entirely on the contents of your own windrow and the environmental conditions where you live. I imagine that someone in the rainier Pacific Northwest will have less need for added water than someone in southern New Mexico. Likewise, if your windrow is placed somewhere that is more shady it will need less water than a windrow more exposed to the elements. The idea is to make sure that your worms have enough moisture to live without having so much moisture that the oxygen is displaced.

If you decided on a trench instead of a windrow and you lined your trench with something impermeable like plastic, then you will want to be careful with the watering to make sure you are not over-watering and causing a puddle in the bottom (and therefore nasty anaerobic conditions, not to mention drowned worms).

In a windrow, if you are using livestock manures then it will serve as both their bedding and also their food if it has enough animal bedding or leftover hay mixed in. If you are just using leaves and such, then you will want to also add some more worm food like leftover kitchen scraps and the like, which we'll talk about in the food section. Leaves and the like are fibrous and don't generally break down very fast and need to be constantly damp to do so. Adding kitchen scraps means your worms have something else to eat while they are waiting for the leaves to break down. You may certainly add your kitchen scraps to the windrow made of manures, but it is not as crucial.

I will reiterate that it is not a good idea to put green grass clippings all over or through your windrow or trench. They will get too hot. Even if your adult worms are able to escape to a cooler area or leave the windrow altogether when it is too hot, the babies and young worms will not. They'll die. Which really throws a wrench into the works if you are trying to convert your garbage into black gold. You want as high a worm population as your worm home can maintain. You won't get that if you keep killing the kids.

I talked about how to cover either a trench or a windrow to maintain moisture and to allow for air flow when I first talked about windrows. There are many ways to do this besides old carpeting; though few are as cheap and readily available. Be forewarned: some of the smaller worms will get stuck in the carpet fibers. Please don't freak yourself out about this, it is only a few and generally they will find their way back out to where they need to be. They are up in the carpet because they are eating the adhesives and the pieces of old food you dropped in the carpet before it became worm cover. If you don't want to use old carpet, you can use your imagination and whatever is available to you to keep your windrow or trench covered, as long as it allows for proper air flow and moisture retention and does not cause the windrow to be overly heated in summer or get too cold in the winter. It is easier to flood a trench, but also easier to maintain a stable temperature in a trench because it is underground. A windrow is virtually impossible to flood, but it can be a bit harder to maintain proper temperatures in areas with extreme seasonal changes. You can, depending on circumstances, leave your windrow uncovered. I do not recommend this though due only in part to evaporation. The real problem is when the neighborhood birds find out what you have been hiding in that big pile of poop. Composting worms are too expensive to use as bird food.

Worm Chow:

FOOD GLORIOUS FOOD!

Yeah! You made it to the food part! I don't know why, but this is the part about raising worms that my students and friends seem the most curious about and interested in. It is also the area that brings me the greatest laughs when I read what other worm growers say the worms will or will not eat. I have an idea where a lot of this misinformation comes from, but sometimes the mind boggles when I read about the types of things some folks tell you to "*never ever never ever never*" feed your worms. The main things I want you to know and remember about worm food are, in a nutshell, these:

  -excessive salts and oils must be avoided
  -there *is* such a thing as "too much" food
  -worms eat any and all decomposing organic matter,
    they do not eat things that are not organic   (again,
    by "organic" I don't mean grown   without
    pesticides, I mean stuff that used to be living)
  -remember that some things can get very hot when they
    are decomposing

First I'll cover the food itself and then we'll get into how to gather and store it, how to prepare it and how to serve it and

89

how much to serve (sort of like you're a chef in a wee worm restaurant!).

Good Worm Chow vs. Bad Worm Chow

Generally, any food from your kitchen that you are not going to eat can go to the worms, oily and salty things excepted. Some foods will get eaten up much faster than others. This is because some foods take way longer to decompose than others and also because the worms do seem to have some "preferences". They dig things like watermelon rinds and coffee grounds way more than onions and orange peels. But just because I like good dark chocolate way more than I like broccoli, doesn't mean I won't eat both and the same holds true for your worms. Though like me, they'll eat the yummier stuff first (pretty sure that is one of the best things about being a grown up: I can eat my dessert *before* dinner if I want). This, I think, is where most of the misconceptions about the things you should "never" feed your worms come from.

Either someone fed way too much of one kind of not-so-favorite food to their worms (like, say, all the leftover pulp and peels from making fresh lemonade) and then found the worms would not eat it as fast as their "regular" food or they had some "objectionable" food in with the worm equivalent of dark chocolate and then decided that the worms *could* not eat a certain food since they were not touching it in the presence of "dessert". Sometimes, what people see and what is actually going on are not the same things. Your worms WILL eat all kinds of decomposing organic matter. Period. The exceptions of oil and salts are exceptions because these things, when fed all by themselves or in excess, can be harmful to your worms, not because the foods they don't eat first are harmful for them.

When I say not to give your worms oil, what I mean is if you get out your fryer and make up a passel of fried chicken or french fries, the leftover oil will need to find a place to go besides your worm bin (or windrow or anywhere there are worms). What a great time to find your local biodiesel aficionado (again with the Craig's List) as anyone who makes or uses biodiesel is going to be very happy with your used oil. Remember that the worms need to be moist because they breathe through their skin. While oil *is* moist, it is the wrong kind of moist. Too much oil will coat the worms, making it hard or impossible for them to breathe. No breath = no life. That said, if you have a few pieces of fried zucchini left over, it won't kill your worms to put those pieces in with their food. Don't make their whole weekly meal out of leftover fried food (or yours for that matter, but this is not that kind of book), a little bit, though, isn't going to hurt the worms.

Alternatively, if you have something that is naturally oily, like nuts or avocados, go right ahead and give them to your worms. Worms *will* eat oil and fat. They just won't live if their environment is not allowing them to breath. They like to nibble on it at will, not wallow in it. Nuts take longer to decompose because they are hard and naturally low in moisture and so are slow to decompose. But your worms will also be happy to eat peanut butter or avocados and similar foods. If you have a lot of this (like four rotten avocados, for instance) don't give this much all at once in an average sized bin and when you do feed it to them, mix the avocado with a little bedding to increase the surface area that will be exposed to the "elements" and therefore will be broken down faster. The worms will chomp down rotten avocado mixed with bedding way faster than just putting a halved avocado in the bin and way faster than just throwing the whole thing in there uncut. And yes, they will eat the skin and even the pit, too. These just take much longer to decompose so it will *feel* like they are not eating it. What they can't eat is that stupid little sticker on the outside of the peel because that is made of some kind of plastic. (Does anyone need a sticker to tell you it's an avocado?)

Salt. Salt is everywhere. Seems to be anyway. It is so hard to have leftovers without a little salt on them. At least at my house that's the case. Nutritional recommendations from your own doctor aside, you will want to avoid *overly* salty things for your worm bin. If you salt a dish while you're cooking it and even when you are eating, it's no big deal. Give the leftovers to your worms if you like. Even most leftovers from processed, packaged food do not have enough salt in them to qualify as overly salty as far as your worms are concerned. What does qualify as too salty is the leftover bowl of popcorn you burned or the dish you over salted so much even you can't eat it or the chip crumbs from the bottom of the bag or, most commonly, the pickle or olive "juice" (brine) left in the jar. These things will be way too salty to feed the worms. If you want to keep worms to eat your organic garbage, sometimes it can feel sooooo hard to throw this stuff away. I totally understand. Salty foods like pickle brine, however, DO go in the garbage. Force yourself. There is no good place to put it. Maybe you could invent one.

There are two reasons to not give overly salty things to your worms. For one, worms are made mostly of water. Salt sucks water out of everything. Worms are a part of "everything". No water in your worms means no worms. But the main thing about overly salty food is that there is no swift and easy way to get excess salt out of soil. Worm poo included. Too much salt is not good for the vast majority of plants you will be interested in growing. Once it is in the soil, the only way to get it "out" is to water excessively and leach it out. Leaching salt out of soil just means diluting it and spreading it around, it is not actually going away. Remember: there is no *"away"*. So, hard as it is, put your overly salty food remnants in the trash and maybe fund a grant program to help the few soil scientists there are out there find a way to deal with this.

Okay, so **no** excessive amounts of oil or salt, but what else? I'll say this very clearly yet again: ***worms can eat any other decomposed organic matter.*** And now I'll tell you why I have to put that in italics *and* boldface! (Don't make me break out the all caps!)

One of the main reasons I decided to write this book was because of all the borderline hysteria I found on the internet about worms both when I was first researching before my own first worm endeavor and after I had been doing it awhile and wanted to find some kindred souls in the worm world. (Yes, raising a lot of worms does tend to separate one from the casual conversationalists of the world. Sigh.) I say "borderline" hysteria only because I'm trying to be polite. I could go on and on with examples, but I doubt those people would recognize themselves and, frankly, this book is funny enough without them. I'll give you a hint though: any time someone is writing about their worms and they repeatedly refer to them as "my WORMIES!!!!!!!" complete with multiple exclamation points: run. Run far, far away. I'm not saying they are never right about anything that has to do with worms, but I will tell you that they are so incredibly *sincere* in their wrong thinking that it's hard not to believe them. Your best weapons of defense: common sense and a wee bit o' logic. Only your common sense and ability to think logically will keep you from falling for these urban worm legends (Wormban legends? Think it will catch on?).

Most urban worm legends are about what the worms can eat all the time and what you definitely can never ever nevereverunderanycircumstances never ever give them to eat. Here is the larger part of the "never" list off the top of my head: oranges, lemons, limes, bananas, pineapple, hot peppers, cheese, milk, chicken, beef, pork, fish, beer, wine, onions and garlic. Here is a list of things I routinely give my worms to eat: oranges, lemons, limes, bananas, pineapple, hot peppers, cheese, milk, beer, wine, onions and garlic. I was a vegetarian for twenty years and now generally only eat meat I can trace the source of so there is not a lot of leftover meat around this house (please don't tell my Labrador retriever that there is *any*). But I have given them beef, chicken, antelope, deer, elk and pork and even shrimp and crab shells and fish and bones. *They eat it all.* And quite happily.

Oranges, lemons, limes and pineapple seem to be a no-no to people because of the acidity. Ironically, I've never heard a single caveat regarding tomatoes, which are very acidic. I used to think that you should probably not give a ton of citrus, but having given literally buckets of rotten tomatoes to my worms with zero ill effects, I can't even say that anymore.

Technically, citrus fruits are slightly more acidic than tomatoes are. Most of what I have read says, with all sorts of dire warnings attached, to not feed *any* citrus. Even though citrus fruits are more acidic, I'm pretty sure the buckets of tomato pieces I have left over from getting them ready for winter storage would trump a single orange or two and still: no problems. And yes, I've given them all sorts of citrus, both the rinds and the whole fruits. Whole entire rotten citrus fruits. No dead worms. Hmmmmmm.

I think part of the problem that caused this legend to get started in the first place and then perpetuated is that citrus *rinds* take a very long time to decompose. I think this may have let some folks believe that the worms won't have anything to do with the citrus and therefore it must be bad for them. But I also think most people have more rinds as leftovers than they do the insides of the fruits. The "acidity" issue is probably just from common sense gone awry. Tomatoes are almost as acidic as most citrus, and there are usually more leftover parts of an acidic tomato than there are of any citrus fruit. But I never hear about tomatoes being bad for your worms. Why? Because they aren't. And neither is citrus.

If for some reason you do have large pieces of peeling or whole fruit, then yes, it will take longer than other foods to get eaten, but not because your worms won't eat it or because it is bad for them, but merely because they eat things that are *decomposed* and the thicker rinds take longer to decompose. Whole fruits take even longer unless you make a point to rip them into pieces or at the very least poke a hole in the side so the magical bacteria can start doing their job. More on how to make food decompose faster in a minute and more on those wonderful, magical bacteria and such, too.

Bananas. The wormban legends surrounding bananas actually slay me. I roll around on the floor, laughing off parts of me that are completely unaffected by aerobic exercises every time I read about the banana problem. Actually, I wish people would spread this particular rumor more often: I might drop a size if they did. Here is the banana story, or to be more accurate: the banana peel story, as told by countless people whom I do hope prefer to remain anonymous: Don't ever feed bananas to your worms because bananas have *so much* *pesticide* on the peeling that even **_one_** could kill **_your whole_** **_worm bin!!!!!!!!!!_**

No, really, that's what they say. Let's put that one through the logic filter, shall we?

Okay, yes, bananas are not grown in the Midwest region of the United States, right next to the soybeans and corn. They have to be flown in from the kinds of places most of us long to be when we are on vacation. Places not necessarily following the "standards" or food "laws" of this country (Look, I have to put those in quotes or else digress here to a soapbox rant which would take up the rest of this book. If you want to know why I put that in quotes, please refer to any number of actually related but non-worm books such as "The Omnivore's Dilemma" by Michael Pollan or movies like "Food, Inc." or "King Corn".) So the theory is based, I guess, on the idea that these countries are so very culturally backwards and bananas are grown by such swarthy and dastardly budding capitalists that they are sprayed with incredible amounts of pesticides to remove all the hairy tarantulas we all "know" would otherwise be lurking in our bananas. No, no one ever says herbicides or fungicides, which I would imagine would be more of a concern in the areas of the world where bananas are grown. It is always the pesticides that seem to matter (and yes, pesticides *can* but not necessarily *will* hurt your worms, but I'll save the rest of that for when we talk about using the poop in your garden).

Here would be the part where you think I'm going to tell you all about food importation laws and how bananas are really grown and how safe they really are. Sorry, I don't even eat bananas. The smell actually makes me a wee bit ill. So I have zero interest in looking that kind of stuff up for you. Sorry. Google it. Or: use logic.

Who besides, I'm sure you will say, a bunch of athletes, eat the most bananas in this country? Short people you say? Why yes! As a matter of fact you would be correct. Short people otherwise known as toddlers and other small humans who make you get the heebies when you are standing in the line with them for rows fifteen through twenty-five on a non-stop five hour flight. They are banana eating fools. And who, might I ask, is in a group of people (or people-in-training) who are also quite susceptible to catching all sorts of bad bugs and whatnot due to their immune-systems-in-training? My logic filter tells me that if just **one** banana peel would kill my **_whole worm bin!!!!!!!!!!!!_** then likely there would be a toddler or two getting at least a tummy ache from the "massive" amounts of pesticides sprayed all over every banana coming into this fine country. Right? Maybe not the itty bitty ones since they can't peel their own bananas, but definitely the slightly bigger short people who insist they can do it themselves (foot stomp and scrunched up face inserted here). And maybe even the itty bitty ones too since Mommy and Daddy don't usually wash up after peeling but before feeding.

But wait. That is not happening. Has not, in fact, happened. Period. I mean really, how many toddlers getting sick would it really take before bananas became worse than cigarettes? Two? Three? The point is that this whole bananas-as-poison idea is just nuts. Plain and simple. Putting it through the logic filter just causes a clog because it can barely get past the opening. Bananas don't harm your worms any more than they harm your little kiddos.

How in the world would some silly rumor like this get started and then repeated so much that people think it is fact? Hmmmm, this one is not quite so easy as citrus. I *think* that one or two beginning wormers who also eat bananas may have royally screwed up and done one of the things you really should never do (like deprive them of oxygen or drown them). Then when they found their worms all dead and smelly, they saw that all or most of whatever other food they put in there was gone or mostly gone except for the banana peel and then used only their hysterical "logic" (my wormies!!!!!) to decide the only remaining suspect, the banana, must be the cause. Poor banana doesn't even have arms and still holds the smoking gun. Almost makes me feel sorry for the banana. But not enough to eat 'em.

At first, I fed my worms only organic bananas 'cause I fell for this myth my ownself. Shame on me. Then, I put it through the logic filter and decided to tell my worm food peeps (the folks who save their kitchen garbage for my worms...Is that love or what?) to go ahead and give me the bananas that were not organically grown. You know what happened? Nothing but poo. Worm poo, of course. Banana peels take longer than other kitchen trash to decompose, than say, green beans or cantaloupe. So it would make sense that other food would get eaten before the banana peel would. They even get lots of actual whole bananas since the people who would previously save rotten bananas in the freezer because they were going to make banana bread "someday" are entirely too happy to cross that "someday" off their list and give the rotten bananas to my worms. If you get these whole bananas, make sure to pull a piece of the peel back or stick a hole in it with your thumb to make things break down faster or it will take awhile before the worms will get to the inside. But when they do...Boy howdy! Do they love them bananas!

Hot peppers, onions and garlic generally fall into the same category of hysteria as being "too strong" for worms. Apparently the people who are anthropomorphizing their "wormies" are also spreading crazy rumors. I say: stop it. Right now. Yes, I realize that for some poor, poor people in this world, onions and garlic and hot peppers cause a teensy bit of, ahem, let's just say intestinal distress. That does not mean that the same holds true for worms.

Worms don't "taste" food like we do. They don't have tongues or taste buds. I mean really, would God give taste buds to any creature designed to eat only stuff so rotted it is soup? Of course not. So the things that might cause us to gasp and cry out for something cold and wet like beer to take the pain away don't cause the same reaction in worms. (Though they do like beer. A lot. And wine. Maybe they aren't entirely stupid after all. Which means you *can* give them your dregs, but only the dregs. Thou shalt not waste beer. Or wine.)

"Hot" food has as much affect on them as "cool" food does. I think this rumor started because peppers are tough skinned and onions and whole garlic take longer than other, smaller and more watery veggies to decompose because of their multiple skins. Leave a whole fresh pepper, seeds and all, in the bin for the worms and it will take almost as long as an orange peel to decompose if you don't poke a hole in it or rip it up so the insides are exposed.

Same with onions and garlic, you have to break it down a bit for the worms or live with them not eating it straight off. If you are a cook then you know that often an onion will be "rotten" on the outside but will be perfectly fine a few layers down. They just take awhile to get gushy all the way through and gushy is what it takes to be worm food. Plus, that papery skin that seems so delicate and breaks off in a zillion pieces all over your counter? It is actually pretty tough and hard for the worms to get through on their own because it doesn't break down fast. The good news is that for most onions and garlic bulbs gone bad, you will normally have cut or broken into them a bit in order to find that they are rotten all the way through. That is great for the worms as it increases the surface area and that in turn causes it to decompose faster. Since you already have the knife out and it is already gross with rotten onion, slice it a couple or five more times before you put it in the worm food container. Then you will find that they eat onions just fine.

I do eat onions. I LOVE onions. And garlic. Life is just not as fine without garlic. Hot peppers I grew into. I grew up in Iowa. State food motto: I'll take mine bland, please. I didn't know what hot peppers were until I moved to Colorado. Every state in the West and Southwest has its own version of green chili. If you have never had green chili, or worse never heard of it, *find it.* Green chili done right is the spicy version of Nirvana. Heaven on earth. Green chili is made from roasted hot peppers. You roast the chilies over fire, then discard the stems and skins and seeds. The stems and skins and seeds, if you are one of my worm food peeps, get tossed en masse into the worm food bucket. This happens only in the late summer, when the hot pepper harvest comes in. And when I say "comes in" what I mean is that IT COMES IN.

I mentioned "buckets of tomatoes" before. This is the same thing but with hot peppers of every kind. The first time I received the blessing of pepper leavings for my worms, I freaked out a bit and was very worried about giving the worms my peppers. I hadn't yet realized that the "my wormies!!!!!" people were so crazy. Um, I mean: I hadn't realized that the "my wormies!!!!!" people needed so much daily medication...

I freeze my worm food before I give it to them. I do this both because I get so much rotten food from so many people that freezing is really the only option to keep the neighbors from being inundated with garbage smells and the accompanying flies and also for reasons I'll explain in a minute. Because it is frozen, I usually take it out the night before feeding day so it is all thawed out and I don't freeze my little fingers off. As such, I have no idea what is in the containers and bags until it is thawed out; it's not like I paw through the rotten food scraps when I get them. So when I first received the massive amounts of green chili detritus I now know comes to me from late August through September, I had a zillion worms to feed and what turned out to be gobs and gobs of hot peppers. I panicked at first. "Oh no! I can't feed this to my worms, it will *burn them!*"

But then I realized I didn't have anything else thawed out and had other things to do that day, so I fed it to them. All of it. Lots. And lots. Of very hot pepper skins and stems and seeds and pepper pieces and such left over from someone making a big ol' batch of green chili. What happened to all my "wormies"? Well, they ate it. All of it. 'Cept the skins and stems. Those take longer to decompose and are seemingly not affected by freezing. It took a couple of more weeks, but all the skins and such got ate, too. That's how I found out for sure that the "hot" in hot peppers doesn't matter to the worms one whit. Same goes for onions and garlic, though I don't feed them whole onions only (without other food to occupy the worms while the onion is decomposing) and never without freezing or chopping first because the onions will actually start growing if there is any non-rotten root left. But the main point is: they *will* eat it.

Dairy, meats and fish I'll just lump all in one category. There is really only one legitimate reason to not feed these to your worms: they can stink to high heaven if there is a lot of either rotting in one place. I know, I know, you have heard that worms actually *won't* eat meat and/or that these things will attract all kinds of nasty vermin (they never say what vermin precisely, but I'm pretty sure they mean rats). I hate to do this so often, but let's bring out our common sense again. Don't be afraid. It's okay to bring it out into the light, common sense *likes* the light.

I'll deal with the smell issue in a sec, right now let's start with the "won't eat it" legend. Worms eat decomposed organic matter. Animals of all kinds, including you, are organic matter. When an animal dies, the decomposition process starts. Eventually, any organic matter with enough moisture present will get so decomposed that it essentially becomes soup. This soup is also known for our purposes as worm food. Yeah, yeah, I know it's "gross". But not nearly as gross as all of that organic matter filling up every landfill and then some. I'll explain more about how worms eat soon enough, but trust me, it is not as gross as your mind is making it out to be.

As I mentioned, I get food from all sorts of people. Everyone gets a list with all the foods the worms like and all the things they don't. Salts and oils are, of course, on the no-no list. I also include meat and too much dairy only because I don't want a ton of everyone else's rotting chicken bones. It is merely a matter of volume for me at this point. But sometimes I get some of the 'no' things on the list. Salted peanut shells during baseball season are a big one. Once, when I was first starting out, I also got an entire chicken leg. Thigh, skin, meat, bones and all. What is a worm girl to do? You already know: I gave it to them. Actually, I put the whole thing in one bin, covered it up really well (the 'why' of covering in a bit), didn't give them anything else to eat and nothing else but water for two weeks. There weren't a ton of worms in this bin, so a whole chicken leg and thigh was quite a bit of food. At this point in my worm growing I was only beginning to realize that so much of the "knowledge" about feeding worms was totally wrong. If it was going to cause a problem, I didn't want to lose a lot of worms.

So what did I have in two weeks? A bone. The femur actually. Everything else, including all the other bones, were gone. Turned from a nasty rotting chicken leg into delicious and delightful worm poop. It was a big "duh!" moment for me, especially considering I already knew that some dairies and farms compost their downed animals, both with and without added worms, with great success. No, I don't know why I couldn't put two and two together. I was obviously on the brink of worm hysteria (my wormies!!!!!) but thankfully logic reached out its lovely, lovely hand and pulled me back in.

The other common legend surrounding the use of meat scraps and dairy in a worm bin is in regard to vermin. Definitely rats, perhaps some raccoons, or any animal that might be interested in rooting in your garbage (bears, anyone?). But here's the deal with that: for one, there are not a lot of animals that eat disgusting rotten meat that don't have "Labrador retriever" in their name. Also, most of the types of animals that would give you concern with meat in your bin would also be happy rooting through your bin for the *other* types of food scraps and are not going to be in there *only* if there is meat. Most animals, like people, prefer fresh food. If they are garbage eaters they are going to eat all kinds of garbage, not just meat and cheese. Sorry, no spaghetti western vultures are going to be flying in lazy circles over your house, just waiting for your old, moldy meatloaf.

Secondly, whether this is a potential problem depends totally on where your bin is and the environment you live in. The vast majority of geographical areas in the U.S. are not normally rat infested. If you don't normally have rats hanging around your place, it is very unlikely that you will attract them from miles away by throwing an old chicken bone in your bin.

Particularly since you are going to be burying your food and most people keep the bin inside. If your bin is outside, the food is still buried or covered and this is not going to be a problem any more than it will be indoors unless you live in an area where these types of garbage eating creatures are common, like raccoons and bears and rats. In which case you will already know about them and will naturally take precautions to make your bin garbage-eatin'-critter-proof. The same way you would already be doing for an outside trash can. If critters aren't messing with food scraps in your trash can or compost pile, they won't mysteriously appear just because you now have worms eating your scraps instead. For most people, your own dog is going to be your biggest critter worry regardless of what you feed your worms. If you keep your worms outside, neighborhood robins and the like will be your biggest source of concern (trust me on that one, those robins are wily little feathered beasts).

If you do decide to add meat or dairy products, I would caution you to not add gobs and gobs at once to an indoor bin if it is very fatty. If you do have more than a handful, just make sure you mix it in with the bedding or poop (depending on the stage your bin is in) to spread it out a bit to increase the speed of decomposition. Depending on the amount, prepare yourself for a wee bit of a very unpleasant smell. But only if you uncover the food before it has a chance to decompose and get eaten. You won't be feeding your worms a ton of food at once anyway, and it is unlikely that you will have a lot leftover meat to give them. Most folks just have a little bit, if any. That stuff is expensive! A little bit, though, is of zero concern for problems or smells. If you have more than a little bit (like more than a serving), it's best if you don't add it all at once unless your bin is at least a third converted to worm poop. Worm poop has an amazing ability to mitigate icky smells from rotting food.

Go ahead and put bones in your bin, but if you have only one small bin, be prepared for the ones larger than your little finger to take quite awhile to break down on their own (except fish bones, which are generally delicate and break down rather quickly). Bones are one of the few organic things that don't turn to soup when they decompose. They do add valuable minerals, though, and some grit for your worms, which they need. So feel free to add a few and please don't fret. I like to put the bones in a bin and let the worms pick them clean, then I pull them out and bury them in my garden. Free bone meal (calcium and other minerals) for my plants. Plus, it will add mystery to your story if archeologists ever excavate your garden.

Since we were just talking about vermin, I will tell you that few things attract mice (vermin if there ever was vermin!) like an accessible worm home. I didn't have mice when I had my worms in the basement, likely because I harbor felines in my humble domicile. However, when I moved them out to the breezeway, they found my worm bins to be a super-cool place to live: it was warm, there were no predators, and there was a handy buffet available 24/7. I have to smack the outside of my breezeway worm bins and outdoor windrows with a broom handle to get the mice out before I feed. But also understand that I am surrounded by fields and farms, so I am going to have lots of mice around anyway. You won't suddenly get mice in your kitchen bin if you don't have them already where you live. If you get mice in your bin, a little peanut butter in a trap will take care of them rather quickly. Live traps are plenty available if the thought of dispatching a mouse does not appeal to you the same way it appeals to me. Look, I'm not *afraid* of mice, I just don't like them much okay? Especially when they scamper over my feet. Geez.

I can't think of any other myths that I've heard about. Your worms can eat all of the leftovers you have for them. The LOVE any watery food (like melon) best, but just because they eat that first doesn't mean they won't eat or don't like the other stuff. I remember when I first started out I read one man's post on a worm forum about how he gets old bread regularly from a nearby store. He said how much his worms loved the bread and that it was *the* major component of their diet. I remember thinking that either I was doing something wrong or else I had some way different worms than that guy because bread was the always the last thing to go in my bin. I finally learned to make sure I soaked any bread or similar dry carbs to make sure they were very soft (mushy-soft) and now they like it better, though bread is still not what my worms go for first.

That about covers the don'ts and the wormban legends surrounding worm chow. How about some more information on the feeding itself?

No matter how hard you try not to, there will be waste generated from the food you eat. Um...I mean the "*before* you eat it" kind of waste, smart aleck. Peelings, stems, seeds, egg shells, moldy bread, stale cereal...Okay, okay, even leftovers from your dinner plate (Don't you know there are kids starving in <u>insert appropriate country here</u>?)...All would normally go either in the garbage or in your regular compost pile. No more! Now these things are all worm food. Remember: worm poop is much more valuable to your plants than regular compost so it is worth what some people consider to be the "extra" effort to make it. I don't think it is extra effort at all, quite to the contrary in fact. But it does involve a change in the way you might normally deal with your food scraps so it *feels* like extra effort. But only at first and only because it is something you aren't used to yet.

All of your non-oily, non-salty food scraps are suitable for your worms, especially egg shells. Worms have a gizzard, just like a chicken, which helps them to break down the food they eat in much the same way your teeth do it for you. A gizzard doesn't work properly without some grit. If you don't eat eggs often, either ask someone who does to save their shells for you or throw in a small (*small*) handful of sand or crushed rock or crushed shells (the sea kind, not the pasta kind) and mix it in well when you make your bin. If you use sand or crushed rock, you only need to add it at the beginning each time you make a new bin. Sand taken from a salt water beach should be rinsed a couple of times with water before you add it to your bin to get rid of the excess salt. Egg shells and a certain amount of the crushed sea shells will eventually break down completely so if this is your grit source then you will want to add them regularly. Same holds true for packaged bone meal (available in a garden center near you for less than ten bucks), which is a substitute for egg shells. But you know what I said about buying things for your worms, so try to find some free garbage instead. The important thing is to make sure there is always some kind of grit available for your worms.

There are people who would advise you to add lime in place of crushed rock or sand or the like. The problem is that there are lots of different kinds of lime and some are fine for the worms and some are very harmful and since it can be confusing once you get to the store to buy it and since there are many available alternatives, and since, really, enough with the buying stuff for your worms already, I am going to pretend that I never mentioned lime at all.

Even though almost all of your food scraps will be suitable for worm chow, you will find that there are some things that your worms have a huge preference for. Mostly, you can bet that any of your more watery foods (like strawberries and melons and cucumbers) will be a huge hit. But they also love coffee grounds and tea leaves. Boy do they! I don't know why, maybe they are caffeine addicts, too. Tea bags are hard for them to get into, so I always rip mine in half or put at least a small tear in them so the worms can get to the insides. They will eat the tea bag (Unless it is one of those stupid new pyramid shaped bags, which are made out of some kind of nylon or something else equally non-biodegradable; if these are what you use, switch to a different brand. Pretty please.) and they will even eat your coffee filters, too, those just take a little bit longer to get eaten up than the coffee and tea leaves themselves. They love other watery fruits like grapes and plums, but because fruit skins are somewhat tough (they are tough enough to protect the seeds through winter, so just because you wouldn't use them for building material doesn't mean they aren't too tough for your toothless worms to get through) you will want to make sure you smoosh the fruit or otherwise crush them in your fingers so the worms can get to the yummy insides. The skins, and even the pits, will eventually get eaten, too.

You may find that your worms are also...Well, to not put too fine of a face on it: they are drunkards. I'd say they were alcoholics, but they never go to the meetings. How do I know this? Well, besides watery foods and coffee grounds, you will find that when you mix food with yeast in it, like bread, with things that have sugar in it, like fruit, your worms will chomp it down faster than you can practically blink. Yeast and fruit juice will give you a crude form of alcohol. I would caution you about making this concoction too often because it can generate a wee bit of heat, but it is good for a bit of a chuckle every now and then. No, the worms don't act drunk (like you could tell?), but it makes me chuckle that I made them some hootch. Your worms will also appreciate it if you have a dinner party and there is some leftover wine or beer to go ahead and use these dregs as the moisture component when you are doing your bin maintenance (more on maintenance in general coming soon). But seriously, I don't think it is a good idea to make this a habit. I don't know if worms are physically capable of having an addiction, but I do think it is best to error on the side of moderation in regard to this aspect for your indoor bins. Mostly because of the potential heat generation. That said, there are also some progressive wineries and breweries that provide their waste products to worm farms as feed. I do think that, in these cases, the worms become accustomed to that as a main source of food, I'm not sure that they should really have it all the time in a confined space. I'm not a Prohibitionist, just cautious about some things.

Too Much of a Good Thing?

When I say that "almost all of your food scraps are suitable for your worms" I *don't* mean you should actually give them *all* of your food scraps. "What???" you say now. "What is she talking about now? She must be a special brand of crazy." While that could be true, you should know that "all of your food scraps" is not the same thing for everyone. If you are married with two kids and you are all vegetarians, you will likely have a lot more food scraps than the family where nobody cooks much and everyone eats frozen dinners and take out. And then there are all the folks in between and even beyond these two extremes. (Perhaps you might reconsider the "special brand of crazy" comment, eh? Turns out I may actually be trying to cover all the bases for ya so you don't have to deal with any nasty surprises or have any burning, unanswered worm questions. No worries, that's what I'm here for.) There are a few different options when it comes to how your worms are fed and how much they are fed. What method you follow will depend on how many and what type of food scraps *you* usually have.

First off, how *much* you feed depends entirely on how many worms you have. We will talk a little more about how many worms are suitable for your household in the part about buying your first worms. The general rule is that you want to feed your worms their **weight** in food about every five to seven days. In warmer seasons it will be closer to every five days and when it is colder it will more likely be seven to even ten days, depending on how cold it is where they live (if they are indoors all year round, then expect to feed every five to seven days).

You will read a lot on the 'net about how worms will either eat their weight or half their weight in food every day. I'm just going to ignore this huge disparity (50% is a huge disparity to me) and address the food amount vs. the worm weight in general and let you do the math...

You add bedding to a bin and keep it moist not only to give your worms a place to crawl around, but to also give them food. They eat the bedding. All of it. The bedding you will use will be pretty innocuous to your worms (which is why we don't make bedding out of green grass), it gives them somewhere to go where they can get away from the pile of people food you gave them in case there is something there they don't like or if they want to take a "break" or maybe to get very, very personal with another worm (oh, yeah, more on *that* later for sure); but the bedding also provides your worms with some chow.

So when you are figuring out how much food to give your worms, you will want to keep in mind both the kind of food and how much bedding is in your bin. Watery food decomposes, and therefore becomes worm food, more quickly than other types of food scraps like, say rotten potatoes. You will have less bedding as your worms eat it away. At some point you will have mostly worm poop with little to no bedding, so by then you will give more actual food. If you are taking care of your worms well, all along they are going to be making worm whoopee and laying eggs and making more worms. Except for maybe at the very beginning and maybe at the very end when the bin is harvested, you will have nothing more than a good guess as to what the actual weight of your worms is. It's not like you are going to be able to actually count them. (And don't even think about trying to give them names.) The point being, that "feeding them their weight in food" at almost any given time is a bunch o' guess work because for a large part of the bin life they will have bedding to eat as part of their food rations and at the same time, the population is (if you're doing it right) increasing. How much it is increasing depends in part on how well you are taking care of their needs and if you had any hiccups along the way ("hiccups" being a nice way of saying "how often you did something so very wrong that it killed some of your worms and how many of them died when you "hiccupped"). So try not to be too in love with this "your worms will eat their weight in food in X number of days" idea.

That said, since most of you will start with a pound or two of worms, using this weight as a beginning measurement for their food is a great place to *start*. Most American kitchens don't seem to have a food scale. I don't think you really need one. Again, please don't go out and purchase something for your worms if you don't have to. This is so not brain surgery. Look in your cabinet or pantry for something packaged that says it weighs a pound. Most worms are sold by the pound. Find a pound of food (rice, popcorn, barley, cheese, whatever). Make sure you look for a pound of something solid (designated as one "lb.", one pound or 16 ounces) and not sixteen ounces of something liquid. Liquid ounces are a measure of volume, not of weight. Take it in your hand and get used to how it feels. It doesn't feel like a whole lot, does it? That is what a pound feels like and that is what you want to think about when you are feeding your worms. Ballpark, people, don't get crazy trying to get everything to weigh out exactly or you will become that irritating lady in front of you at the deli counter.

Lots of you already save your kitchen scraps for a regular compost pile. In my circle, the people who do this save the scraps in a coffee can or similarly sized container that is usually kept by the sink for this very purpose. This is fine, but remember that most coffee cans are going to hold way more than a pound of food scraps. There is one thing I like to do to get around having to use a "too big" coffee can, and you are certainly not required to do this even though it *is* a great idea. I like to save drink cups from the coffee shop I frequent (and I do mean "frequent"!) and use these to save my food scraps. I have a certain addiction to iced Americanos from a major coffee chain that serves coffee in a size that rhymes with "spendy". This sized cup coincidentally also holds just about a pound of food scraps, give or take of course depending on the type of food scraps (No, this is not a "which weighs more, a pound of feathers or a pound of bricks?" type of measurement riddle. Some of your food scraps will weigh more, like leftover cooked pasta, and some will weigh less, like spinach gone bad.). If you have a similar beverage addiction, this is a great way to reuse these cups at least once before they go to the recycler. The food scraps you generate that go beyond this amount can either be put in the regular compost pile or you can store them for later use (how to store them in a minute). After all, if you are doing everything right, no matter how many worms you start with, eventually you will have more. More worms will need correspondingly more food; so you will be happy to have some extras in storage since it is not likely your food consumption will change much.

Most vermicomposting instructionals will tell you to just put your food scraps right into the worm bin as you make them. While there isn't really anything "wrong" with this, I do think there is a better way.

If you put your scraps in as is, they are usually fresh scraps and can take a bit of time to decompose. The faster you can make the food scraps decompose, the faster they will turn from mere food scraps to actual worm food. The more watery the food, like lettuce and strawberries, the faster the decomposition process even when these types of scraps are fresh. But this also means that the actual food (the "not-just-water" parts) will be somewhat negligible because of the water content. Think of how big a dehydrated strawberry is in relation to a fresh one and you will have a great idea of just what I mean. Alternatively, things like orange rinds and banana peels have less water and more fiber and so are going to decompose more slowly, but when they do the amount of actual worm food is greater than for watery food scraps. Then there are things like cabbage, beets, raw potatoes, raw onions and carrots that also seem to take forever and a day to decompose and often will even start growing again in the bin. No, seriously. We will talk about things growing in your bin later, but do you really want plants growing in your bin?

I've fed this way before; just putting scraps in the bin as I make them. I've also seen other bins that are treated this way. They look pretty sloppy and I always would look in the bin and think "when is this stuff going to get *eaten* already?" At first thought, this may seem the answer to doing things the lazy way, which I've mentioned is my favorite way. But I have found that this actually seems to make the whole process take even longer. Being *effectively* lazy means to use less energy to achieve the same results. But really being good at being lazy, and I am, means to not only be effective and get the same results, but to do it *in less time*. Ohhhh, is that not the coolest idea? Wouldn't you like to know how to do that? Let's learn...

How you deal with your worms' food is going to depend a lot on what you have as scraps to feed them. The rest of the formula will depend on how much of a life outside of your worms you want to have and on how dedicated you are to being lazy.

If your diet and/or lifestyle only gives you simple scraps like coffee grounds, egg shells and banana peels, then go ahead and put everything right into the bin as you get it. Anything you might do to "prepare" these types of scraps will have little overall effect and won't make it worth the effort. However, if you eat a varied diet (and I do hope you do) and therefore have all sorts of things to put in your bin, then you have several options as to how to use these scraps to feed your worms.

One thing you can do to speed the process if you do just want to throw the scraps right in the bin is to chop them up really well before you put them in there. Most folks have the knife and cutting board out anyway, getting their food ready to make whatever for dinner. While everything is cooking you can just pile up the scraps and chopchopchop them all up as small as time allows. This is also a really fun way to leave the frustrations of your day behind. What could be better therapy than chop-it-into-little-bits-therapy? Pretend the vegetable scraps are that quarterly report your boss keeps breathing down your neck about. Just don't get so carried away that you forget about your own dinner.

Alternatively, you can put them in the blender or food processor and push those buttons until everything is one big pile of mush. Not quite as much fun as chopping, but still gets the job done. Be careful if you do this, however, that you don't add too much, if any, extra water. Some is usually fine, but doing this all the time will likely cause your bin to be too moist and will encourage stinky conditions.

There is a problem, however, with fresh food that too many people seem to gloss over: fruit flies. I'll cover this more in the bug section, but if you have ever kept a compost can on your counter you will know what I am talking about. Decaying fresh fruit is the perfect breeding ground for those annoying little fruit flies. They *always* seem to fly right in your face and totally will get into your margarita mix if you forgot to screw the cap on all the way (if you are making margaritas this can have a nasty, party-pooper side effect). There're no two ways about it: fruit flies are a total pain, even if they don't technically hurt you. There are ways to place the food in the bin to lessen the chances of getting fruit flies (we'll get there) but even better there are also ways to make sure you never get them in the first place.

Some folks recommend actually cooking your scraps, either on the stove top or in the microwave. This does, in fact, help the food to break down faster and will, it's true, eliminate the possibility of fruit flies. It also uses energy (both yours and electricity) that doesn't need to be used. Not to mention the impact it has on your social life. Please, if you do decide to use the cooking method to prepare your worm food, *don't tell anyone.* Ever. Unless, of course, you prefer social isolation. In which case telling people that you do this will likely get you what you want.

Your lazy girl author has an even better way. Those "spendy" cups I mentioned using to hold my scraps? First off, make sure you save the lid and keep the lid on the cup when you are not putting scraps in it. Second, once the cup is full but before you feed it to the worms, put it in the freezer for at least overnight. But wait, you say, trying to call me out, this uses energy too! Actually, not really. The size cup I'm talking about is not big enough to cause your freezer to come on and so the cup contents will freeze without using any more energy than you were already using. Best of all, though, is that this freezing will both help the food to decompose faster *and* will kill any fruit fly eggs that happen to be on your scraps. You can totally even do this if you want to use a coffee can for your scraps. In fact, I have found it much more beneficial to have people just keep the can in the freezer from the get go. Especially in the summer when fruit flies can be a bigger problem in your compost can then they will ever be in your worm bin.

Since you are hopefully going to be on a regular schedule for worm feeding, all you have to do is take the cup out of the freezer the night before or morning of "worm food day" and by the time you are ready to feed, the food will slide right out. A perfect portion of practically already decomposed worm food. For most of the food scraps you will have, especially the scraps of watery food, the act of freezing and thawing helps to burst the cell walls, causing the food to start getting mushy (which is the first stage in the decomposition of vegetable matter; "mushy" being the scientific term, of course). Think of the difference in texture between a fresh carrot and a thawed carrot that has been frozen and you will totally understand.

Hey! You say. This seems like a lot more work than just putting the scraps right in the bin, I thought you said this was the lazy girl way? And it *is*. If you want to put your fresh food scraps in every time you make them *and* avoid fruit flies (somewhat) and stinkiness, you have to get in the bin, make a hole, bury the scraps, cover them up, go to the sink, wash your hands, dry them, go back and wash them again when you see you still have a bit of worm poop under your nails...and *then* you can go finish your dinner. I'm not even going to discuss all the time you will waste complaining about the fruit flies that will still grow in the bin and explaining to your spouse how great worm poop is and can you please keep the bin even though there are fruit flies in the margarita mix? Now, doesn't my way seem a bit easier? Plus, in summer or if you live where it is hot most of the time, you don't even have to let the food thaw out completely. You just have to let it thaw enough to get it out of the container. It can finish thawing right in the bin and won't hurt your worms a bit. Talk about easy.

There are also some things you can feed your worms that you may not think of at first as worm food. Most people just think of people food scraps or animal manures and some paper shreds and there ends the list of things to feed the worms. But there are other things that you may also want to use: the contents of your vacumn cleaner bag for one. Floor sweepings. (Unless you have a feisty, big eared, little mutt that likes to rip the guts out of her stuffed toys and strew said toy guts all over the place. Toy guts are usually made of polyester, which is not organic and therefore not worm food. Though please don't tell this to said feisty little mutt, in her mind she really *is* killing the stuffed hot dog.) The stuff in the lint catcher on your clothes dryer is also worm food. The paper towels you use (though remember it is best if you shred these first). The contents of your or your dog or cats hairbrush (Yes, hair is too biodegradable! You've been watching too many B zombie movies and think that it doesn't decompose, but it does. It takes a little longer but it *will* break down and it is my understanding that hair is a great source of nitrogen. Still a doubting Thomas? Put on your logic hat and ponder this: think of all the people and animals that have lived on this planet. Imagine how much hair they would have all had. Now imagine how fuzzy our planet would be if hair didn't decompose. It does.) Actually, your worms will even eventually eat your old clothes if they are made out of something organic like cotton. Great way to get rid of your husband's stinky, hole-ridden "lucky" sweatshirt once and for all, eh? You will probably have more than enough to feed them just sticking to the obvious things, but if you want to reduce your waste output, think outside the box (and inside the bin!).

Okay, now you've got your food in order and you are ready to feed the worms. Yeah! See how far you've come? Just a few pages ago you didn't even have a home for them.

You can absolutely just toss the thawed food into the bin and wait for it to decompose if you like. This is easier to do if you have some kind of lid, especially a lid that rests on top of the bedding like it does in commercially made bins. The problem with this method will be covered in the section on bugs, but hopefully by just saying that you have a good idea why that is not the best practice. Tossing it on top without a proper covering also usually makes your food dry out, which turns it from great worm food into just some dry garbage your worms won't touch.

Instead you definitely want to make a point to bury the food scraps when you feed your worms. This keeps it away from the bad bugs and gets it down in the moist (not wet) bedding and poop so that it will continue to decompose and not just dry out. I have read all sorts of ideas about how to bury the food. This cracks me up since there can't be many things less complicated than digging a little hole, dumping the food in and then covering it back up.

Here are the different food burying plans I have read or heard about: bury it in the very bottom, bury it below the top but not all the way on the bottom, shove aside half of the bedding in the bin and spread the food all out in a thin layer on that side and then cover that back up, switch which sides you do this on whenever you feed, don't switch the sides when you do this, separate your bin into quadrants and then make a chart and then every time you feed, bury the food in a different spot...And on and on. Sounds a little, um...What is the nice word for it...Let's go with 'compulsive'.

How about this: every time you feed, always bury the food *almost* at the bottom and always in the center of the bin. This works well for a few reasons. For one, it's simple and won't make you nuts. Secondly, I don't like to put the food on the very bottom because of the previously mentioned gravity issue. Your food will naturally be somewhat moist on its own, gravity will make the bottom of your bin more moist than the top. Sometimes these two factors will combine and make conditions a teensy bit smelly if the food is at the very bottom. Putting the food a couple of inches from the bottom seems to virtually eliminate this problem. I also like to mix the food up with a little of the bedding or the poop that is in the bin. I'll explain why in a little bit.

The almost most important reason to always feed in the same place is so that you can monitor your worms food intake and therefore readily know if you are feeding too much (because the next time you go to feed and find that there is a lot of the previous feeding left over) or feeding the right amount (because almost all of the food is gone...You do want a little bit left over; you don't want your worms to try to go out for dinner.), or you are not feeding enough, because all of the food is long gone and you can hear all your worms' tiny little tummies grumbling (just kidding about the grumbling part, they don't really have grumbly tummies). But you don't want them to be hungry because if they can't find food they won't be growing and won't be reproducing and if they are really hungry they will try to go find food somewhere else or die trying. They won't have extra bedding around to eat forever, you know.

All of the other different feeding schemes involve putting your food someplace different every time you feed, even if it is just one side or the other. You hopefully have better things to do than try to remember where you fed your worms the last time. But here is the very most important reason to always feed in the center: you are a geek and you are not able to be quiet about that. Look, it's not an insult, but I know you are a geek because you are sitting there right now reading a *whole book* about how to raise *worms*, of all things. (I mean it when I say don't be insulted; after all, I am the bigger geek who *wrote* a whole book on how to raise worms. Picture me on the bow of a big ship, arms spread wide yelling, "I'm queen of the worms!!!" and then you will understand I'm not insulting you but merely welcoming you to my world.)

All your friends already know you are a geek. You will have your friends over and, knowing they already think you are a geek and being afraid they may be correct in their assessment, you will try to keep the fact that you are now raising worms a secret. Try and try. And fail. Sorry, composting worms are just *too cool* to keep quiet about! At first, you will innocently but perhaps a little too exuberantly exclaim: Wanna see my worms?! Your friends will rapidly decline and then likely suddenly remember a very important appointment with a bottle of shampoo. At some point you will realize that when you ask this question, your friends are immediately thinking that you have a medical problem and fear it may be contagious. They *won't* be eating dinner at your house no matter how good of a cook you are. Believe me when I tell you that you will not have such a horrific reaction from your friends if instead you say: You have to see this! And then just go show them the bin without telling them what it is beforehand. (Why no, I didn't learn this the hard way, why ever would you ask?)

Depending on where you keep your bin, you will have to move it to show it off. This is your grand opportunity to gently educate your friends about what you are doing and why and the super cool benefits you are getting from it. Don't be alarmed if some of your friends are a little creeped out still. Remember that some folks were chased on the playground by some screaming little bully waving a wet, wiggly worm at them. Or a snake. Same difference if you are scared of creepy crawlies. But the rest will be fascinated and might not even know why, just like you may have been the first time you heard about this whole crazy idea. Now, since you had to take your bin out and show it all around, it likely got turned around a little bit. If you always feed in the center, you never have to worry about the bin being all turned around because no matter how many times you take it out or how many of your friends turn it all around when they are checking things out, the center of your bin will *always* be at the center. No charts to consult, no sides to worry about. Just the center.

When you first get your bin, I recommend that for the first two months you error on the side of feeding less than you think you might need, but feeding slightly more frequently than you will once you get the hang of the whole idea. Because of the amount of bedding that will be available as additional food for the worms, you won't have to worry about them starving to death. But since overfeeding can be easy to do and can cause huge problems pretty quick, you want to give yourself a month or two to learn how much is enough and how much is too much. Feeding more frequently will also give you the extra opportunities to check the status of your bin and get used to how things will be in your particular set up. I only think you should do this for the first bin you ever have. That is when you will likely make the most mistakes because, hey, you are just learning. Subsequent new bins won't need as much attention because by then you will have the whole process down pat.

To Infinity and Beyond:

Bin Maintenance

Now, you don't just build your bin and save your scraps and every week dump some food in there and then eventually harvest your bin. Well, okay, technically this will mostly work, but it does take longer to get a bin full of poop and it can cause problems in a hurry. I am lazy, but I am also impatient. Yes, it is a horrible combination. I have, fortunately, found a way to satisfy both of these parts of me in a way that sacrifices neither.

Your bin should have some regular maintenance time to keep conditions appropriate for the life of the bin from beginning to the end. *Not* taking the time to do this is not the lazy way, it's the stupid way. If you don't pay attention to what is going on in your bin, you can very easily wind up with dead worms or stinky conditions or worse: both. It is much more work if you only *respond* to problems once you get them or have to start all over because you killed your worms than it is to avoid these problems completely in the first place. My way involves a tiny bit of time and work for about five or ten minutes every week, but it also means you won't have to deal with dead worms or stinky bins. Definitely a good trade off, trust me there. Plus, my way means your bin will be converted as fast as humanly possible...Oops, I meant as fast as wormanly possible, of course.

However often you are going to be feeding is also going to be how often you do your maintenance. You shouldn't have to get in your bin multiple times to do multiple chores, that's just too much work and we don't do that here. Your bin maintenance should take less than ten minutes a week. Less than five minutes once you get used to the process. Though sometimes it is just nice to sit there for a couple of extra minutes and marvel at this wonderful, sweet process.

First, you will get your food ready and have it handy, but don't put it in just yet. Now, some folks love the feel of their hands in dirt and I happen to be one of them. There is nothing in your bin that will hurt you and once it is about fifty percent turned, it will feel and smell just like you are working in really nice, super soft dirt. That said, I also don't think it is worth it to have to scrub my hands and clean my nails just for five minutes worth of work. As such, I wear rubber gloves when I am working in my bin.

You can decide to wear or to not wear gloves. I've read some tripe in the past that you shouldn't *ever* put your bare hands in your worm bin because "the oils on your hands will kill your worms". No, really, I've seen that more than once. Please understand that this is just b-o-l-o-g-n-a (got the song in your head yet?). For one, there are no oils coming from your hands. You don't have any oil producing glands in your hands. Any oils that might be on your hands are there because you touched something oily, like the dirty dishes or you patted someone else on the head and they had some wicked nasty product in their hair. If you are worried about "the oils from your hands" then just wash them already and be done with that whole idea.

I will tell you this quick story, though: when I was up to about 10 indoor bins, I was starting to accept food from all sorts of places because by then I never had enough food of my own to feed the inside and the outside worms. I'd take just about any rotten food. Someone gave me about eighteen cases of packaged tofu that had gone bad. Cases. Each case had about eight packages, all trying to burst their seams because they were decomposing and getting gaseous in their packages. Up until this point, I just did all my bin maintenance and feeding with bare hands. You can only imagine the unholy smell when I started opening those packages. Oh my. But I thought the tofu would be great for worm chow since it was already going bad and since tofu is naturally already mushy. So I put the tofu in both the indoor bins and in the outdoor windrows and went in the house to clean up.

That unholy smell? *Would not scrub off my skin.* I tried everything, including soaking my already scrubbed-to-redness hands in warm bleach! Ouch. But no change. I must have washed and scrubbed my hands and nails 59 times that day. About mid-day *the next day* the smell was finally gone. It felt like forever. My nose is *very* sensitive. It took about a week for my hands to recover from all the scrubbing. That is the main reason why I wear gloves now: some foods can be more smelly, and apparently long lasting, than others.

Okay, back to the bin maintenance. You've got your food and you have your gloves on, if you like. First dig in the center and see how much food is left. If it is just a little bit of food or none then you will want to feel around the rest of the bin to see how wet things are. If everything feels right, then assess the height of the bedding. Throw some of your pre-moistened bedding right on top to bring the height up to where it needs to be, if needed. Then just take the bottom, which will be more moist than the top, and mix the bottom and the top all up enough so you can no longer tell which was bottom and which was top. Don't get so energetic with the mixing that the contents start flying out of the bin onto the floor, but you can pretty much mix it up pretty quickly. If anything does fly out of the bin, it's best to wait until it dries before you vacuum it up for the easiest clean up. Unless a worm flies out, of course. Those go right back in the bin where they belong.

When you feel around the bin, if things seem overly moist (like more than 2-3 drops would come out if you picked it up and squeezed it—but remember not to actually squeeze some or you will smoosh your worms) then as you mix, add some dry bedding to even out the moisture. Mix it all up well so everything is at a good moisture level. (If your bin is only slightly too wet, the dry bedding will not soak up the extra moisture right away. Just mix it in there and in a couple of hours, all should be at the proper moisture levels. I don't expect you to mix for two hours. Seriously, your arms would fall off. Just mix and then wait a couple of hours.) If your bin is too dry, add some moist bedding and/or some water. No, I can't tell you how much dry bedding or moist bedding or water to add because I'm not there with you and can't feel or see what condition your bin is in. This is why you are going to feed a little less but more often at the very beginning. It will make everything easier for you if you get in there more often and see not only what needs to be done, but how quickly conditions can change in there.

If you look in the middle and see that there is more than about a ½ cup of food left from the last feeding, you will still do the rest of the bin moisture check and mixing, but when you are doing this you will avoid the center area and leave the food where it is. (If there is *a lot* of food left you will also be checking to make sure you still have worms in there!) You want to avoid mixing too much food up into the top parts of the bin to avoid any possible bug problems, but a little bit of food mixed up into the bedding is not going to hurt anything.

Once your bin bedding is mixed up you can feed the worms. Dig your hole in the middle, reach over to the food you already set nearby and dump the food in the hole. See how smart that was? If you didn't have it out ahead of time, you would get worm poop all over the place when you go to get the food out. Or you would have to take off your rubber gloves, get the food, then try to put the gloves back on. Not as easy as one might wish. Those rubber gloves are hard to put back on once you've been wearing them.

One other thing I learned from the "tofu incident" is that sometimes the food scraps will get a little smooshed together and not decompose as rapidly as it should. This happened with the tofu because I put the tofu blocks in the bins "as is". The centers of the blocks went even more anaerobic than they were in their little packages. Boy, I thought it was smelly when I *first* put it in there. Rotten tofu left to get even more rotten in a warm worm bin is likely the smelliest thing I have ever had the displeasure of breathing in. And I have been around some highly stinky things! I learned that if I mix the food up with a little bit of the bedding or poop from the bin before I cover it up, not only does it get eaten faster, but any smelliness the food may have when you put it in is gone in less than a day. So put the food in, mix it up a little bit with the bin contents (be it poop or bedding or any combination of the two) then cover it up and put the lid on. You are DONE. It likely took you longer to read about how to do this then it will take you to actually do it.

Performing this regular maintenance every time you feed your worms will speed up the process and will also give you the opportunity to correct any problems with the moisture or air. It's great to keep your worms healthy, but the cherry part of this system is the "speeding up" part! After awhile, you will get used to the conditions in *your* bin and then the maintenance will only take a minute or two. That is the beauty of it, but also the problem for some of you. When something only takes a teensy bit of time it is also easy to forget. I'd try to make it harder and take more time for you so it might be easier to remember, but I just can't. So many other people have tried to make it harder, I had to write a book to explain how simple it all really is. Sorry. Maybe you could set the alarm on your cell phone to remind you.

Feeding, Watering and Maintenance for Windrows and Trenches

Everything I've said so far about feeding and maintenance to this point applies to a regular sized bin, such as you would keep inside your home. Outside boxes in the garden have generally the same rules if you want things to go fast. Trenches and windrows, however, are far too large to be maintained in the same way as an indoor bin. I mean really, if you have even a small windrow that is say, two feet wide by four feet long by a foot high, it would take you all morning to feed and turn and such. What a drag! We don't do things that aren't fun so if you are going to have a larger worm home, your rules will be a little bit different. They are a lot more work at the beginning and the end, but a lot less work in the middle.

If your trench or windrow has been set up properly you will have a way to water regularly and thoroughly enough that at some point everything will be moist, even if not always at the same time. By that I mean that if you use a soaker hose in a windrow, depending on how many hoses and how large your windrow is, not all of it will be moist all of the time. The area of the windrow closest to the hose will be wet and a few inches out it will be just moist and if you skimped on the hoses, there will be areas that are actually dry. Keeping your windrow covered will help the moisture wick to all areas over time and as the worms eat the area closest to the hose and turn it into moisture-holding poop, it will be even easier to keep the whole thing moist. If you find that you have more than just a few isolated pockets of dryness, make sure that every week or two (every week if it is over about 80 degrees or so), you supplement the moisture by spraying the hose over everything and make sure you do a better job of hose placement the next time.

Presuming the water is properly set up, feeding is the next thing to think about. Some people like to bury the food even when it is outside. To me, that depends mostly on what you are covering the windrow or trench with. As I mentioned before, I use old carpet. Despite the porosity of carpet, it still does a great job of holding in moisture and keeping out the things you don't want in the windrow (critters and unwelcome bugs). I weigh down the corners and some of the long edges for the larger windrows with cement blocks or bricks. As such, when it is time for me to feed the outside worms, I just pull back the carpet and put the food right on top. The rotten food juices will help add moisture, just like in a regular bin, so keep this in mind since the scale is going to be exponentially larger. You can certainly bury the food if you like, but if you are using soaker hoses this can be a pain. You will be most familiar with how you have your windrow set up and can make the best determination on the "where" of feeding for your own system. If you have a system or an idea for a system that involves you having to bury the food, you may want to think of ways to alter the type of covering to allow you to just put the food on top and save yourself and your back from all the extra digging.

I do recommend that you still freeze all the food before you feed in your larger scale worm home. The reasons are all the same, but so much more important when you consider the difference in the amounts of food you will be giving. It may seem like a pain, but you can get a really nice freezer for cheap and often even for free from Craig's List and assorted used appliance stores. I certainly do not recommend buying a brand new freezer to store garbage in. You can reduce your energy consumption when the freezer is not full by keeping some jugs or soda bottles filled about three-quarters full of water near your freezer and filling the freezer with these when it is not full of worm food. A full freezer uses a lot less energy than an empty one.

I feed my outside worms only about every three weeks except for in the summer, when I feed about every ten days to two weeks. I know this is not how most other folks do it. Most folks feed their outdoor worms much more often. There is nothing wrong with that. Commercial worm farms, those far larger than mine will likely ever be, feed differently too. I don't follow their schedule since my worms eat a variety of miscellaneous garbage and most worm farms feed a steady supply of either homemade or a commercially available worm chow (yep, there sure is such a thing). They do this to avoid problems like having several cases of rotten tofu to contend with, but mostly because they have the time/feed/worms ratios down pat and throwing in "unknowns" is not always profitable. I want my worms to eat garbage, period. I'm willing to accept the "risks" and a slightly less predictable time frame of completion in order to have such a tremendous impact on the waste stream. I think it's a good trade off. I may lose a few worms from time to time or have stinky food for them, but I also get to be responsible for keeping literally *tons* of organic waste out of the landfill. I WIN! (And you do too.)

I feed on such a different time frame for my outside worms for three reasons. The most obvious reason is that there are a lot more worms out there and a lot larger area to put the food in (or on). The scale is larger so the amount of food is larger than in an indoor bin. Not, however, proportionately more, but about twice as much. The second reason is because, frankly, it is a pain. Old wet carpet is heavy and I have a lot of outside worms. It takes awhile to get everything that needs to be done accomplished. I'll reiterate: if I don't have to spend energy on something, I won't. Better Homes and Gardens™ is never going to knock on my door asking to photograph my lovely worm windrows. Guaranteed. So I'm not going to put in extra work making them "perfect". Or pretty. They are definitely not pretty. Lucky for me, my worms don't have eyeballs. And though they are happy to eat a shredded Better Homes and Gardens™, they aren't going to read it either.

The third (and main) reason I can feed much more but less often in my outdoor windrows than I can in an indoor bin is specifically because of the limited environment of an indoor bin in relation to an outside windrow or trench. In an indoor bin, as I mentioned, if you overfeed and conditions go bad as a result, your worms really don't have many places they can go to get away (and live) until conditions right themselves. In an large outdoor situation and by feeding at the top (or even if you bury the food along the top, but not all over), the worms have lots of places to go to get away from any bad conditions that might be present in the area where there is too much food. For one, there is a lot more "down" in an outside windrow than there will be in an indoor bin (No, you can't just make your indoor bin deeper. We covered that. Go back and read that part again and then you can finish this part. Go on. Go.) The worms can not only go down if they need to, they can go to the edges and, if your windrow is straight on the ground, they can go there, too. Don't fret, they will come back to where the food is.

Just like when we talked about putting green grass on your windrow, putting down this much food at once is also not something you do without a plan. Lots of rotten food can cause some nastiness. Worms like rotting food, but too much of a good thing can be too much for them to handle, just like it is for you. When I put the food down, I lay the food down pretty thick for a couple of feet, then leave about a foot or two with little or no new food in between the areas with food. This helps keep the worms where the food is, where they will be eating at the edges of the food even if the center of the food is too thick or hot or otherwise icky for the worms. Remember, the worms will be eating the manure or whatever is the major component of the bedding in addition to the new food. But you also want to give new food regularly to help feed the ever increasing worm population and to keep your freezer from overflowing.

A couple of other things should be mentioned here in regard to maintaining your bin, be it indoors or out. The crazy people have reared their heads about many things and one of them they go on and on about is using a pH meter to check the pH of your bin to make sure it is not too acidic or too alkaline. Worms like a more neutral pH just like you do. But they can live a little above or below just fine. The problem with a pH meter is that it doesn't come with worm instructions. Depending on where you put the meter, you will get different readings. Stick it right in the middle of a bunch of rotten tomatoes you just threw in the bin for food and you will get a very different reading than taking a reading from the area near the edge where there is mostly shredded paper or poop. I suppose you could take readings from, say, ten different places, write them all down and take an average.

But just by writing that anyone would ever consider doing that is making my head explode. I bought a pH meter when I first started. Hey, that's what the book said to do! Actually, I spent twelve bucks cold cash on a combination moisture, light and pH meter. Fancy. I used it a couple of times when I remembered I had it. And then one day when I was cleaning out the garage, I noticed it laying there, covered in dust. I also noticed that my worms were just fine and, in fact, thriving without it. So I sold it on Craig's List for ten bucks.

The thing about the pH and the way I am teaching you to raise your worms is that doing it my way means that *if* there are some bad conditions in your bin they will more likely be at the very bottom (from gravity acting on the water) or in the food pile. Which means there are lots of other places in your bin that your worms can go to get away from bad conditions. Which means they won't die. When you do your regular mixing, that will likely take care of any of these bad conditions and then your worms will be able to squirm around wherever they want to. *Poof* the pH meter problem takes care of itself. And you just saved yourself twelve bucks.

There is one thing that the vermicompost naysayers will tell you about worm poop that is actually true: the downside of vermicomposting is that it does not take care of seeds the way hot composting does. Hot composting properly (Remember how much work that is?) will destroy or kill almost all seeds so that they don't sprout later when you use the compost in your garden. Vermicomposting does not do this. Some seeds will certainly be destroyed, but the very nature of seeds also means that some won't. The seed hulls are *designed* to protect the seeds. But! There are some upsides that don't often get mentioned.

For one, if you make it a habit to freeze the food before you feed it to your worms, you take care of a lot of the problem right there. A lot of the food you eat has seeds associated with it that are not designed to withstand harsh temperatures. Secondly, some of the seeds that can withstand freezing will actually sprout right there in your bin. It's actually pretty funny since there is little to no light for them. You will remove the cover and there will be a sprout or six, all pale and yellow winding their way to wherever they think they are going. All you need do is pluck them and throw them back into the bin, where they will disintegrate and never be a problem again. Some of the seeds will simply decompose, just like the rest of your food. Not every seed is a potential plant, not even in the package you buy them in.

Lastly, most of the things you use as worm food will be scraps from food you like to eat. Having some "volunteers" in the garden from the unsprouted seeds in my vermicompost (No, you won't be able to see the seeds and pick them out, sorry; they are small and will be covered in worm poop and will just look like pieces of worm poop themselves.) has never been a problem. Actually, I've had some of my very best tomatoes from volunteers I've allowed to grow because they were not in a bad spot and weren't competing with any other plants. Really, this is *not* going to be a problem for you. Unless...Unless you want to vermicompost your weeds. Weeds that have *not* gone to seed can absolutely be fed to your worms (pre-composted of course, if they are green). Weeds that have gone to seed should not.

Look, I know you want to be lazy like me so you may not weed your garden quite as often as you probably should. Some weeds, only one or two I'm sure, may have escaped your diligent attention and reached maturity. You can certainly put these weeds in your hot compost pile, if you have one, to destroy the seeds and *then* feed it to your worms (not a bad idea since the worms will make your regular compost even more delectable to your plants). Or you can remove and throw away the seed head and vermicompost the rest of the plant. Or, because that seems pretty high maintenance to me, you can throw the whole thing in the trash. Oh, stop it. Yes it is too okay to throw some things that have surpassed their useful life into the actual garbage. As a gardener, I can hardly think of a better place for weed seeds than in the garbage.

I've gone on and on about how to raise worms in a way that is easy enough to not take up your whole day (ever) and to give you time to have a life. If you are very successful at the having a life part, there are likely going to be times when you are away from your house for extended periods of time. These times are otherwise known as: vacation. Sweet, sweet vacation. Get yourself as many worms as I have and "vacation" will be a mere memory. But for you and your one or three home bins, vacation is definitely in the picture. Maybe your worms want a break from you, too. If you are going away for only a week or ten days, just make sure that the worms are fed their regular amount and watered the day or so before you leave and they should be just as fine while you are gone for a week or so as they would have been if you had been home and not fed them in the same amount of time.

If, however, you are a lucky one and get to go somewhere fun for longer than that, like two or three weeks or more, you may want to make provisions for your worms. "Making provisions" does **not**, and I can hardly emphasize this enough, mean giving two or three weeks worth of food to your worms at once. I already went over the dangers of over feeding and feeding more than a week or so worth of food at once is definitely over feeding. Your worms don't know you are going on vacation, even if you tell them, no matter how slowly or loudly you talk. The same holds true for giving a bunch of extra water. We already covered those problems so I won't go over them all again except to say: no. No, no, no.

What should you do then, if you are going to be gone for a long time? That answer, like so many others, depends. If you are reading this book and planning to start your first bin about a month before you go on vacation, my recommendation to you is to wait to start your bin until you get back. Or if you just harvested and started a new bin, same problem. The major part of your bin maintenance is always going to be needed at the beginning when paper and such makes up the most of your bin. Remember, the paper doesn't hold the moisture in the bin like the poop will. So if you don't plan well or have to go away unexpectedly for more than a week, you will definitely want to make sure you get a worm-sitter to, at the very least, make sure your worms have enough moisture. They will have bedding to eat, so you needn't worry too much about the food part if it is just a couple of weeks. But if it is longer than that, make up a few weekly "servings" of food for the sitter to feed.

If, however, your bin is anywhere between a third and two-thirds finished, then likely the moisture issue will take care of itself for a couple or three weeks as long as you have a cover on your bin. If not, at least put a piece of cardboard over half of the top to slow evaporation. Same story on the food. Though if you are going to be gone longer than that, at least have someone come in every couple of weeks to give some food and a little moisture, even if they don't do the maintenance part (and if you can find a worm-sitter to do the maintenance part, count yourself as very blessed and make sure you bring back a souvenir better than an airport T-shirt).

But what to do if your bin is ready to harvest or very close? Well, you certainly don't want to go ahead and harvest and then start a new bin. If you do then you get to jump back a couple of paragraphs and read about that problem all over again. Instead, all you need to do is add more moist bedding to give your worms something extra to eat once they eat up all the actual food. Though, yes, the same moisture and "no extra food" rules still apply. Essentially the rule is that your indoor worms will be fine for a week or two without you, but longer, you need to make sure that you have someone else to make sure they don't die while you are off in Morocco.

So that's it for the food. Nope, I'm not going to give you a long list of all the foods the worms *do* eat. Only the don't list (excessive salt or oils) is enough. Everything else is fair game as long as it came from an organic (carbon based) life form.

Section Two:

The Worms
(and Their Friends)

-WoW is really World of Worms
(types of worms)

-population control
(how many)

-worm sex
(PG-13)

-bugs, molds and bacteria, oh my!
(their friends)

The Worms

"Finally," you say, "we are to the actual worms! Why didn't she just start there?" Well, duh. If I started there you would have gotten all excited and went out and got your worms and then forgotten to read the part about how to raise them until it was too late. No, I haven't met you before. Just lots and lots of people like you. It's truly great that you are so excited to start. Hopefully by now you understand there is a wee bit more to this worm composting business than just throwing them into a pile of food scraps.

First we will talk about the different kinds of worms you might encounter, then about how many you will want for the type of worm home you are going to have. After that, we get to the PG-13 section and learn about how worms make babies. Then, we will talk about the invited and the uninvited house guests you can expect to find living (mostly peaceably) with your worms. All of it is important to know, so let's get started.

Wigglers, Crawlers and Their Handlers

Depending on where you are getting your information, there are at least three to four THOUSAND different varieties of earthworms sharing this great planet with us. I have also read numbers as high as six and seven thousand, but since most of what I read from credible sources ballparks the numbers as between three and four thousand, that is what we'll go with. The only reason you need to know that is for a trivia contest, it has almost no relevance to your vermicomposting.

I'll tell you now that there are very, very few people who study worms scientifically. We literally know more about the workings of the human brain than we know about this most important and prevalent of God's creatures. It's sad, but part of the problem is that I'm guessing "studying worms" doesn't sound as sexy or interesting as "studying rocket science" does to a twenty-year-old college student. Or to the young woman he's trying to pick up. Doesn't matter that there are so few rockets and so many more worms to study. Probably the same reason we don't have many proctologists, who also have an abundance of material to study versus rockets. (If you are thinking that studying worms sounds cool but are worried about how that might potentially affect your dating life, try telling them something like, "I'm studying how to convert unused organic matter into an earth-friendly, sustainable and far superior soil enhancer." Way better than "I study worms." Trust me on this.)

What I'm telling you is that, based on my research, there is some contradictory information out there both in the popular vermicomposting literature and in the hoity-toity soil science journals. It would be super-cool if this were not true, but I'm going to leave sorting all that out to the ones who wear white coats because most of it doesn't really matter as far as growing worms at home is concerned. I promise I'll let you know what I know and give you alternative theories if they are relevant.

One thing that *does* have relevance to your vermicomposting is how the many different worms are broken down into different types. There are three general types of earthworms, only one of which you can use to make poop in a controlled environment so you can have direct poop access.

Endogeic worms live down in the soil and are generally not the ones you ever see unless you are digging a deep hole to plant a tree or the like. Endogeic means "within the earth" and that's where they like to stay. They eat certain types of organic matter and minerals already present in the soil, they don't feast on your kitchen scraps.

Anecic (which means "out of the earth") worms are the same ones you may have hunted in the wee hours of dawn before you went fishing and before you got a job and found out you could just sleep in and go buy them a whole heck of a lot quicker. Anecic worms also live down in the soil, but they come up to the top to get some chow and to leave their waste. But they don't hang out on the top. Most people commonly call these "night crawlers", but not all of the worms called night crawlers will fit into this category and not every worm in this category has "night crawler" in its common name, so don't confuse a common name with any actual indicator of the type of worm.

Both anecic and endogeic worms are loners. They don't like to hang out much with other worms except for the occasional romp to make baby worms. They also like living in the soil. By "in the soil", I mean lots and lots of soil, lest you get the idea that you can raise these worms inside if only you fill a big box with dirt. Each type also will make permanent burrows to live in, which they can't do in a box with a bunch of other worms. Both types like freedom of movement even more than they don't like living with other worms. So: you can't use these worms for your bin.

It is unlikely that you will get these types of worms from a commercial worm grower (the person you will get worms from either in person or over the internet) or even from a bait shop (more on growers and bait shops in a minute). If these types of worms don't like to live with other worms in your house, they aren't going to do so for a grower or be very happy at the bait shop. (As if any worm is going to be happy sitting around waiting to be bait.) You will, however, frequently find these worms sold at little stands in areas of the country where kids still go worm hunting in the dark hours for a little extra money during their summer breaks. By all means, buy those worms for fishing. But not for your bin.

Epigeic worms are your goal. Epigeic means "upon the earth" and these are the kinds of worms that live on the earth's surface or just a couple three inches down. They feed on decomposing organic matter and thoroughly enjoy each others' company. They also don't need to travel much, making them ideal for living in a controlled environment.

So where would you get these kinds of worms? Well, don't google "epigeic", that's for sure. But you can google "composting worms" if you like. That is what they are most commonly called. They also go by many other names, the most common of which is "red wiggler". Red wigglers and all the worms that are sold under this name and similar names, are smaller worms that are great for composting. They are usually more red in color than pink and about as big around as a standard piece of yarn and about 3-4 inches long when mature. They are great composters and very easy keepers.

There is a long, long list of common names for the different types of worms you can use for composting. Some of them will have the same name and be the same worm, some will have the same name but really be different worms. Did you know there isn't one source to go to for information on worm taxonomy? You can find information on every different real and made up type of dog breed all over the place. But try to find out what kind of worm you really have in your hand and good luck! To add to the confusion, a lot of worms will look alike to the naked eye, and only be discernable from one another microscopically. If you had a microscope. And if you knew what you were looking for.

The good news is that you don't *need* to know what kind of worm you have, as long as you have worms sold as composting worms. Lots of places will tell you they sell "red wigglers" or even name them as Eisenia foetida or Eisenia andrei or the like. But unless they sent a handful to a worm taxonomist (now there's someone not usually found in the yellow pages) to find out what kind of worms they are really selling, you will never know for sure.

Most people who raise or sell worms say they have such-and-such type of worm because that is what the person they bought them from said they were and so on and so on. If you go to someplace online like findworms.com, you will think there are a LOT of worm growers out there. But the reality is that the commercial worm growing world is pretty small and quite a few of those growers started out with worms from one of the other growers so essentially they are all just telling the same thing they were told.

True, just because I've never heard of someone actually finding a worm taxonomist and sending their worms in to one doesn't mean someone didn't do it. And if they did? It still doesn't matter. It only matters that the worms you are growing are the kind that will live in a bin and eat your kitchen scraps. It's just like when you send in your dogs' DNA to have it analyzed to find out what kind of mutt you really have. It doesn't matter that you now know you have a Newfie-weimeriner-poodle-beagle-corgi mix. It only matters that your dog does what dogs do. Any worm sold as a "wiggler" or Eisenia foetida will do what you want them to do. They are smaller worms that eat like madmen and reproduce quite well and make you all kinds of happy with how your composting efforts turn out.

Unless you like to fish. Then you may want to explore the world of larger composters. If you like to fish and already bought a "red wiggler" type of smaller worm, you can find all sorts of "worm fattener" recipes online and you can even find a few places that will sell you "worm fattening" worm chow. I say, DON'T DO IT! If you want to fatten up your red wigglers for fishing, you can certainly do it. But it does involve getting some foods like milk powder and chicken layer feed that you would likely not otherwise either buy or have as food waste. Please don't buy food for your worms. If you want fatter worms for fishing then *just buy a bigger type of worm* (but first try the littler worms for fishing, you may be surprised at how successful you are).

There are lots of larger kinds of composting worms most often sold under the name of "something or other" night crawler. The most common name for big, fat composting-type worms is "European night crawler". Sometimes this is also "Canadian night crawler" but it is my understanding that a true Canadian night crawler is anecic and therefore not a good composter. I've purchased worms labeled as "European night crawler" (or ENC for short) from different growers and gotten completely different kinds of worms. I mean worms that were obviously visually different; no need for me to pull out my trusty ol' microscope. I've even purchased ENC's from the same grower and received a completely different worm from the kind I'd gotten before. (This I think was a case of re-selling, which you want to avoid and which I'll get to in a second.) In any case, you can find composting worms with "night crawler" in their name that will fit the bill both for composting and for the fatter worms you would like for fishing. They are great composters, just like the little ones, you will just get fewer per pound because they are larger.

Composting (as opposed to fishing) worms sold with the word "night crawler" in their name are not going to be true night crawlers. Meaning they are not anecic or endogeic but are actually epigeic worms that are just fat enough to be sold as "night crawlers" to people who fish. Make sense? If not, no worries, it only matters that you purchase worms sold for composting.

But which kind of worm is the best? Don't be so silly, okay? Everyone always wants to know which kind is "the best". Sheesh. If I don't tell my customers and then tell them why I won't tell them, they will buy both kinds I sell so they can find out for themselves. No one has ever come back to tell me how much better one is than the other or how much faster or slower one worm seems to be, etc. You know why? Because *there is little difference.* A composting worm is a composting worm. Period. You are going to raise them to eat your garbage, not to solve quadratic equations, right? They will all do the same thing for you and the only difference that will matter to you is if you want fat worms to fish with. I promise.

I personally prefer the larger worms only because they are easier to see and therefore easier to harvest. It is my experience that the smaller composting worms eat more quickly but are also more delicate by nature. The larger composting worms don't seem to turn a bin into fabulous worm poop quite as fast, but do seem to be able to handle a larger range of poor conditions than the little guys can.

I do also know that answer will not satisfy you so I'll tell you this about buying your worms: get your first worms from an established grower or from a friend who has successfully been keeping worms and has an excess. Make sure if you get them from an established grower, you get worms that are sold as "composting worms" of some name or another, I don't care what kind. Well, actually I do a little. I've found that places that sell their worms with the very generic and plain "composting worms" tend to have a variety of species in their mix. My personal philosophy is that a mix is always going to be better in the long run. A mix will most closely mimic natural conditions and a more diverse species base will give you a leg up when there are minor problems in the bin that might be major problems for only one of those species.

What I mean is: biodiversity generally gives you advantages that you will hopefully never even notice. For instance, if you have (and for some reason can be sure you have) all of one type of worm, and if your worms get sick somehow or there is something highly objectionable in the bin then it will likely affect *all* of your worms. If you have lots of different types of worms in your bin and something bad comes along that affects one kind, you will still have all the other worms in your bin that are not affected and that will continue to thrive. Make sense? I hope so. Biodiversity matters in all natural situations and your worm bin should not be an exception.

The good news is that most of the growers that sell anything called "red wigglers" or something similar likely already have a mix of types and may not even realize it. Most likely their mix will not contain any of the fatter varieties of worms or night crawlers you may want for fishing, though, so keep that in mind if you want fat worms. A few farms sell their worms as the very generic "composting worms" with no distinctions made as to type. These growers generally sell a mix of the larger night crawlers and the smaller wigglers.

Most of the worms sold as night crawlers, though, do tend to be all of the same type, but that has much more to do with the type of business that sells big, fat worms to fishermen and women as opposed to the type of business that is only geared towards composting worms. Fishermen are statistically 207% more superstitious than even baseball players and are generally the type to both notice a change in the type of worm and also blame that minor change on why they had to bring home fish from the grocery store instead of from the lake.

I definitely recommend that you try to find worms from a grower that is close to you geographically; at least within your temperate region. You can get worms shipped to you from all parts of the country. No matter where you are, it is more likely than not that you will have to have your first worms shipped to you regardless; like I said, there just aren't that many of us. But you want to try to find someone relatively near to you, at least within a couple of states and at least for your first worms, to make sure that you are getting worms that are suitable for your environment.

Not all worms sold as composting worms can live everywhere. Some growers in the far southern regions of the U.S. have types of worms that are not suited to ever live in the more northern regions and vice versa, even if they might be okay in your house. Some do, but some don't. This is fine while they are in your house, but if you are going to use their poop, which will have worm eggs (cocoons) in it, in your garden, you will want those worms to be able to hatch and thrive in the conditions available to them outside *your* home. So make sure you get worms from someone relatively nearby and who raises their worms outdoors rather than in a controlled environment in order to give you the best chance of success from beginning to end.

"Nearby" in this case does not mean the bait shop. Most bait shops do have worms suitable for composting, but they are generally on their last legs by the time you get them (after all, they are not destined for a long life, right?) and they may or may not recover once you get them into your bin. If you really feel this is your only alternative, try to find out what day they get their fresh shipments in and only get your worms on that day. But worms at the bait shop are wickedly expensive, so even if you factor in shipping and handling for composting worms, you are likely paying more per pound than you would be if you just bought them from a composting worm farm. (Or worm ranch; people have their own preferences for what they call their business. I choose farm instead of ranch only because I don't have enough room in my barn for all the saddles.) If you have your heart set on it though, you may check the labels on the container the worms come in at the bait shop. You may be surprised to find that they come from a grower who is relatively close to you and who may be willing to sell you some worms straight from their business (minus the shipping and time passed and refrigeration) instead of through the bait shop.

At the time this book is being written, composting worms are being sold for between twenty and thirty dollars a pound, plus shipping. Holy wow! Really? Yes, really. I do realize this is more expensive than steak, but then again, how many times has your steak made more steak for you? (Um, I mean *after* it has been turned into steak, smarty.) I do realize that most folks are surprised at how much worms can cost initially (aren't you glad I talked you out of buying things for your worms now?), but remember that if you are successful, they will make more of themselves and you will not only never have to buy them again, you will likely even have a couple of extra pounds a year and you can sell these yourself on Craig's List or the like and recoup your costs. If that is not good enough, think of this: if you use kitchen trash bags (and I think most people do), your use of these bags will decrease significantly. Most folks don't wait until their kitchen trash bag is filled to capacity because there is usually something rotting in there that must get out of the house *now*. But: raising worms means they will eat that rotten stuff for you so you will not need to take out the trash so often and not need to use so many bags (and if there is something that you don't feed your worms that might get stinky if you put it in the kitchen trash, just keep it in the freezer until trash day, okay?). You get to save money on bags and save the landfills from that much more plastic. Win! I will tell you some more ways your worms will save you money, potentially significantly more money than kitchen trash bags, when we talk about actually using the poop in the next section.

I mentioned before about being aware of worm re-selling. This doesn't happen often, but it does happen and it can be a problem. The main problem is that it can lead to dead worms, and rather quickly after you get them. The second problem is that it is nearly impossible for you to know if you are getting worms from someone who is only buying and re-selling and not raising the worms themselves. This re-selling generally happens only at peak season for commercial worm growers. See, people who want worms generally want them the most in the spring and through about the middle of summer. Worms, worms, I need some worms! I get some seriously desperate calls, you should hear them. First off, you don't "need" worms right this minute. If everyone came to terms with this, I don't think the worm re-selling problem would exist. But they don't, so it does.

Even knowing that they can sell a ton of worms every spring and summer, there is a finite amount of worms that any one grower will have. It is very difficult for a grower to tell a customer that they are sold out and likely that they are sold out for the season or at least for a couple three weeks at a minimum. During these times, demand literally always far outweighs supply.

Worms can only reproduce so fast, regardless of how good the grower is at their job. But no one wants to turn customers away because they think they will never come back. This is not true, but that is what people believe. (If you are a worm grower and reading this: I tell people I am sold out when my population gets to a low point and *still* I have more customers than I can deal with the rest of the year, okay? "Honesty is the best policy" is a cliché for a reason.) So what do they do? If they can find them (and not all growers sell to the public, so they can), growers who are loathe to tell you they are sold out will get as many worms as they can from wherever they can. Most often, the "wherever" part is not someplace they can just drive to one afternoon. Instead, the worms are shipped from one grower to another.

Sometimes, the first grower can order more worms in anticipation of the rush and the worms they get from the second grower will go into their new home with the first grower for at least a couple of weeks before they are shipped out to fill an order. This lets them recuperate a bit from the multiple stresses of being shipped. Sometimes, though, things just go too fast and worms will get shipped from one person and then shipped out from the receiving grower right away. And this is where re-selling becomes a huge problem for anyone receiving them.

Shipping is hard on worms. They are usually shipped with a little bit of moist bedding to keep them alive for the time that shipping takes, but they will always be a little bit dehydrated and hungry by the time you get them. Sometimes shipping can take a few days or even a week. This is a long time for worms to not have adequate moisture or food (remember: they are small and mostly made of water). If a seller has worms shipped to him and then promptly re-sends them without a sufficient recuperation period, it is very difficult for the worms to recover, especially because this happens mostly only during peak season. Peak season always happens in the middle of summer when everywhere is at its hottest. When you get your worms, they may even look bad or possibly even be dead on arrival. Any reputable grower will guarantee that their worms will *arrive* to you healthy and alive, but if you don't know what to look for, you may not know to ask for a refund or re-shipment. Most growers will not refund money for worms that die after arrival. This is because they don't know how good you are at raising worms and have no way to know if you did something to kill them with ineptitude rather than that there was something wrong with their worms. At twenty or thirty bucks a pound plus shipping, this is a lot of hassle and expense for both of you. So if your worms are not moist and wiggly within a couple of hours after you receive them, make sure you contact your grower and discuss your concerns with them right away in case they don't make it into the next couple of days alive. If you wait, you likely will have to pay full price for more worms.

How to avoid this mess? Simple. Make a point to buy your worms at the beginning of the season or at the very end. This would be spring and fall. The best times to buy worms for sure. Since anyone who practices re-selling is unlikely to tell you if they do it or when they do it, this will greatly increase your chances to have healthy worms on arrival. The problem with summer and winter is because sometimes the worms have to be shipped long distances. It is unlikely that your worms will be shipped in anything more insulating than a cardboard box. This means that if it is eighty degrees outside anywhere along the worms route, it will be 100 degrees inside the truck they are being transported in. Likewise when it is very cold outside, it will be that much colder for your worms. Don't even get me started on where your mailbox is located and where your worms will be sitting, baking or freezing, between the time your mail is delivered and the time you get home from work.

Which brings me to the other two important things about buying online: 1) make sure you read the shipper's policies regarding both when and how they ship and 2) make sure you understand what the grower's guarantee policy is (if any) and what the complaint or return procedures are. Most growers ship only one or two days a week. They do this on purpose to take advantage of shipping schedules at the post office so the worms are in transit for the shortest possible time. Most growers will email you the day your worms are shipped and most will tell you an estimated date of delivery. When you are ready to order worms, please remember that it is *your* responsibility to make sure you or someone is either home to receive them or will get home as soon as possible when they arrive to get them out of the extreme temperatures of your mailbox and into their new home.

Any reputable online seller will list their complete return/complaint information on their website. Make sure that you both fully understand and agree with this policy *before* you order and that the website contains contact information beyond merely emailing through the website. No, you shouldn't expect your grower to be available to you twenty-four seven, but having more than one way to contact the grower in the event of a potential problem will give you piece of mind that you are dealing with someone of integrity. They may not be able to get back with you in five minutes, but they will get back with you. It is in the best interest of every reputable grower to make sure you are both happy with their stock and successful in your vermicomposting efforts.

If you buy from someone in your own state or city, but still have them shipped, then you can buy them whenever you like as long as the weather is not going to be extreme between the time you order them and the time they arrive in your mailbox. If the grower is close enough for you to drive to get them yourself, then you can pick up your new worms at any time of the year.

You may have to make a mini-day trip to get your worms if you want to pick them up, but it will be worth it to know they will arrive healthy. Now don't start yelling at me about driving and using gas to get the worms. The delivery truck that would otherwise bring them to you doesn't likely run on purified water. Combine errands to make it less of a single reason trip, which really everyone should be doing anyway, right? Or, just wait a couple of months until it is a good time of year to get them shipped. It's not like getting the worms is an emergency. You are okay with waiting to get tomato plants until the right time of year to plant them. Think of worms like that, they are living things after all, and thinking that way will make it easier to wait until the time is right.

Population Control:

How Many is Enough?

Okay, so you found a grower and figured out the best way to get them to you, but how many should you get? Almost all composting worm growers sell worms by the pound. You can find some people on Craig's List and other similar sites who will sell them in lesser amounts, but a pound is a great place to start. It is commonly stated that there are around a thousand worms in a pound. Now, I must admit, I have never counted them. This may be right, or it may not. I presume that is a good average, but of course the total number will depend on the age of the worms, younger worms are smaller than older worms. This "thousand a pound" number only applies to the smaller composting worms (wigglers), of course, not the night crawlers. The night crawlers are usually at least twice as big as the wiggler-size worms, so expect to not have as many in total number. If you get a pound of worms and want to count them, that's super. Please do let me know what number you come up with!

Having repeated the "thousand in a pound" dogma, I do have to advise that I do not recommend that you order from anyone who sells worms in any configuration other than by weight. The problem with the "thousand in a pound" idea (besides the fact that I have never really counted and am only repeating what I've heard-shame on me) is that *no one* ever counts them. I frequently order from different growers in a quest to find someone I can recommend to my own customers to when I run out of worms to sell. I did, once, order worms from someone who sold them by number. I paid about $25 for "1000 worms". When I received them, they were not in good shape and obviously very dehydrated. I weighed them, bedding and all, and the whole shebang was just over a third of a pound. I was not happy, but besides counting them I had little recourse. I didn't want to count them because they were very lethargic and truly needed to be put into a better environment right away. The worms lived, but I'm not sure they would have if I wasn't as experienced with growing worms. The point being, though, that that "thousand" sure wasn't anywhere near a pound in this case. Actually, they didn't even look like a few hundred. I think these places bank on no one ever actually counting the worms. I didn't, which is my bad and why I never called to complain. But hopefully you will never fall for this scheme.

An actual pound of worms is a great starting place for the average household. Remember that family of four who are all vegetarians I mentioned awhile back? Those folks should start with two pounds. A family of six or more that eats a lot of veggies? Two or maybe even three pounds. I do *not* recommend, for any beginner of *any* sized household that you start out with more than three or four pounds; even if you have grand plans to start your very own commercial worm farm. I really recommend that you start with just a pound, maybe two.

There is some skill to this process and a lot of what I call "green thumb intuition". You will have a better start if you already have a green thumb and a love of dirt and gardening and a nurturing spirit. Not having a green thumb doesn't mean you are doomed to failure by any means. The fact that you are reading this book will give you a great foundation, but nothing beats actually doing it to understand the ins and outs. If you get more than a couple of pounds to start with and then find out the hard way that you are really, really bad at it and as a result, hate it so much you neglect your worms, then what good does that do for anyone? Start small and your mistakes will be small. If you do well, your worms will increase their population for you. For free.

I also absolutely recommend that you take your first bin all the way from beginning to harvest before you decide to expand your efforts. Each stage of growing worms has its own easy parts and hard parts. You may find, for instance, that growing the worms is cake for you but you absolutely cannot *stand* to do the harvesting. If you found that you liked raising the worms so much that you went ahead and started a few more bins before you ever got to the harvesting part and then found out you hated it, you would have that many more bins full of potentially dreadful work ahead of you. Alternatively, if you get from beginning to end successfully (and most of you will) and find that you think this is one of the coolest things around and ask why no one ever told you about this *years ago*...Well, then, by all means get more worms if you like. But if you successfully get from beginning to end, you will have many more worms than you started with anyway, so why rush things and get too many to start with? Oh, that's right, you have lots of extra money. Of course. The worms won't give you quite as great a return as a good mutual fund will though, so a little well-timed patience goes a long way towards both great gardens and a fat wallet.

Remember, too, that more worms also means they will need more food so make sure you can feed them when you need to without having to actually purchase food. (Though you will be amazed at how many people will save their garbage for you if only you ask!) If you only start with a couple of pounds but your house generates more food than they can eat, then just put the excess in your outside compost pile until your worm population increases either naturally or because you really are ready to buy more. The great thing about this whole process is how cool it is and how great worm poop is for your plants. The problem is that sometimes people get a little too crazy about the whole thing and go overboard before they are truly ready to take on that kind of work with the knowledge they have. So please, start with just a regular tub for a bin and a pound or two of worms and work your way up to more if you want or need to. You will thank me for this advice, promise.

I mentioned before that you will want about a pound of worms per cubic foot of bedding in your bin to start. If you made a small bin to start but have your heart set on two pounds of worms, you will want either two bins or one bin large enough for at least a pound and a half of worms (meaning 1.5 cubic feet of bedding total). Yes, I realize these numbers don't add up. Two pounds should be in two cubic feet of bedding, you say. Well, sure. But again, this is vermicomposting. We are not splitting the atom here so total precision is not required. Since your worms will make more worms, a bin built for a pound of worms will become overcrowded far more quickly if you put in two pounds of worms at the start. Most likely it will be overcrowded before it is time to harvest. It is far more convenient to split your worm population at harvest time than it is halfway through. Two pounds will do okay in a home built for a pound and a half, but don't get much smaller than that or you will have worms trying to leave to find a less crowded habitat well before you are ready to make one. If you want to start with three pounds of worms, then you will need three regular sized bins or two slightly larger ones to take a pound and a half each. I'll talk about overcrowding more in a little bit.

The vast majority of growers who sell worms by the pound will sell them as what is called "bed run". For 99% of these growers, bed run means that you will get a pound of worms of varying ages. Some will be big, some will be little, but collectively they will make up a pound of worms. Most often when you get your worms, whether you pick them up or they are shipped, this pound of worms will come with some moist bedding. It is proper for a grower to weigh out a pound of worms and *then* add the bedding and ship them. A small few will have the bedding make up part of the pound. The 1% left over from the previous 99% will just scoop out a pound or so of whatever happens to land in the scoop and sell that and call it a bed run pound. There is no legal definition of a bed run pound of worms so any grower can sell whatever they want to as a "bed run pound". You, as a consumer, will want to make sure how they measure before you give them your money. Obviously, a pound of worms weighed alone, without bedding, is a much better deal for your money. Most reputable growers will tell you right on their website or over the phone just how their worms are weighed.

It is also possible that you will get worms from a friend or from someone on Craig's List or a site like vermicomposters.com. These worms may or may not total a pound. It is likely that if you are getting them from a friend or someone else who has "extras", you will only get a handful or so. A pound of worms is about three or four handfuls, it doesn't seem like much when you get them. But just a handful is not many at all. If you get a smaller amount like this, I highly recommend that you make your first bin about half the size that you would make for a pound of worms. (Half the size of the bin itself, I don't mean you should make your bedding half the depth.)

The problem with a small amount in a regular sized bin is twofold: with so few worms, it will take a much longer time to turn your first bin into worm poop than it would with the right amount of worms. The second problem is that if you have a small number of worms in what to them would be a big bin, they have fewer chances to meet. Fewer chances to meet means fewer chances to mate. Which means your population won't grow very fast and all of these conditions will conspire to make you think it is all just a big waste of time. It's not! I promise.

If you get your worms shipped, remember that this process is not only stressful for them, but they are going to be shipped with only a little bit of moist bedding to keep them alive for their journey. Almost all of the time this is sufficient and your worms will be just fine. But shipping is stressful even under the best of conditions so if you cannot pick up your worms directly from a grower, expect that the worms you get will weigh about a pound *with the bedding* or even a little less when you actually get them. The stress of the journey will likely make them a tiny bit dehydrated and so they will weigh less than when they were shipped. Not significantly less, though, but it is not abnormal for a pound of worms and bedding that weighs just over a pound when shipped will weigh only a pound or just under when it arrives at your door. I only tell you this so you don't feel like you are being cheated on weight, you are not. The most important thing to worry about is if your worms are healthy and can quickly recover from being shipped.

This does not mean that you want to rinse them in water when you get them to "rehydrate" the worms. Remember that too much water is just as bad for them as too little and if you soak them when they get to your house, they'll think they are drowning. That's a lot of misery for worms that just got tumbled around in a box for a few days. You will recall that when we talked about setting up your first bin I said to have it ready with moist bedding and food about a week or ten days before you get your first worms. Hopefully, you did this. If not, hurry up and whip up the bedding and food before you do anything else.

Just take the worms from whatever they were shipped in (usually a bag made out of landscape material or a similar breathable fabric) and dump them "as is" right into your already prepared bin. The worms and the bedding they came in should be dumped in altogether. Make sure there are not any stragglers clinging to the sides of the bag or container. Then: *leave them alone.*

If your worms are healthy, they will know what to do. Come back in twenty minutes and take a peek. All or all but two or six worms should have dived down into the bedding after fifteen or twenty minutes max. If they are all or mostly all still huddled around where you dumped them, wait another twenty minutes just to be on the safe side. If the vast majority have not gone down into the bedding by then, its time to call the grower and let them know there *may* be a problem with the health of the worms. Be patient with the questions the grower asks. It may be a little bit like calling tech support for a computer problem: they don't tell you to turn the computer off and then back on because they think you are stupid, they don't know you and so they need to start at the beginning just like your worm supplier will. They will talk you through what they need you to do in order to get a refund or a new shipment, *if* that turns out to be what you need.

I know that I just told you to leave your worms alone. I also realize that no matter how many times you read this book before you get your worms, there will be a little bit of what can only be described as parental anxiety when you get your first worms and put them in your first bin. It happened to me, too. Actually, it happens to me all the time. I love my worms and want to know they are content. You will want to know your worms are fine, too. It's totally okay, I understand. And because I understand, I will remind you of one of the many myths surrounding worms: it is okay to touch them! They won't die from "the oil on your hands". Of all the ridiculous things.

So when you get the urge to "just check on them for a second" instead of waiting for twenty minutes like I just told you to, understand that it won't hurt them to do this. But also understand that if they are doing what they are made to do, meaning: dive down and avoid the light and get to eating, you won't see them. Which will immediately make you want to do a bed-check and dig around in the bin to see the worms for your ownself. And...you might not. Don't panic. Your worms will blend in *amazingly* well in your paper shreds. Quite the vanishing act. They are in there. Check the corners and near the food, that is where they usually are. Now try to put your anxiety away and leave them alone for awhile, okay? They just had a big journey and want to check out their new digs and rehydrate; which they will do just fine in the moist bedding. Checking on them every five minutes is as annoying to them as that guy at the office is to you when he starts talking to you a mile a minute before you even get your coat off. So try to give them a little time to acclimate.

If your worms do go straight into the bedding when you first put them in there, then everything should be fine and you can leave them alone (except for the aforementioned and apparently irresistible parental bed-check). Until it gets dark. (Insert scary music here. Da-dum. Da-dum. Da-dum.) There are a few times when the disorientation caused by the shipping gets your worms all discombobulated and they will sometimes try to escape the confines of the bin at first. This does not necessarily mean that there is something wrong with your bin, only that it is not the home they are used to. If you check your worms about an hour after sunset and find that a few are crawling up the sides of the bin, or worse, are on the floor outside the bin (don't forget to check underneath the bin, too), scooch the renegades back down into the bedding and, for just the next night or maybe two, put a lamp in the room or put them in the spare bathtub and leave the light on in that room. Since worms are photophobic, they will get the idea and after a day or two and will realize that their new home provides everything they need, even if it is different from what they were previously used to.

But! If after a couple of days they are still trying to escape or if they are trying to escape even with the light on, then you know for sure there is something wrong with the way you assembled your bin. "Trying to escape" includes hanging out all over the sides and lid of the bin, even if none of them actually ever make it out onto the floor. The most likely culprit is that you put far too much food in there to begin with. Take some of it out, at least half, and mix some of the bedding with the remaining food. Make sure there aren't any worms in the food you take out though.

If they are still crawling up the sides, the next most likely problem is that you did not build your bin in a way that allows them to have enough air. Take the lid off and set it on the bin sideways or remove it altogether. This may mean you have to protect it from the cat for a couple of days, but if this does the trick then you know what the problem is and can fix the lid up to allow better airflow and still keep out the cat. The last culprit may be that you put them into some kind of objectionable bedding. If you didn't use something recommended in this book, quick mix something recommended (like shredded paper) and push the old stuff to

153

the side and put the new bedding in half the bin. This will hopefully solve the problem. If not, you have me stumped without knowing what your particular set up is like. Call the grower immediately and see if they can help.

A very small number of growers also sell worm cocoons in addition to selling worms. I'm not sure what the advantage to buying cocoons would be for a home composting situation. I think these are mostly for the people who want to seed their garden or farm with worms in an easier way. But the same holds true for seeding a garden or farm with cocoons as it does for seeding it with worms: if there is not enough for them to eat where you put them, they'll die. Better to build up the organic matter to make sure they have something to eat before you decide to spend your money on cocoons. By then, you should see your native worm population picking up quite nicely and you will have avoided the expense and heartbreak of finding out you put the cart before the horse. I would, in case anyone who sells cocoons is reading this book, be very interested to know how they are harvested. I have a few ideas, but no real idea and think it would be nice to know just for interest. For the rest of you, I'll talk in the harvesting section about the cocoons that will be in your worm poop so hang on, we'll get to it all in due time.

At some point, probably about the time you harvest your first or second bin, you will find out that you have more worms than you can keep up with. Ideally, if you started with one pound of worms you will have pretty close to two pounds of worms by harvesting time of the second harvest, if not the first. You may have some ups and downs learning how this process works with your first bin and so may have only about a pound and a half at the end of your very first vermicomposting endeavor, but subsequent bins should give you pretty close to twice as many worms as when you started. A little before you are ready to harvest, you need to think about what you want to do with these "extra" worms.

Did you just say, "I'll put them in the garden"???? Tell me you did not just say that. You get to go back to the beginning of this book and start over right now. Go. Go on. You can come back to this section when you have a better answer.

You may decide that you want more worm poop and/or that you have enough food or access to food for them and so you want to have more than one bin to hold these "extras". That is wonderful! I'm all for people growing as many worms as they want to if and when they are able to. It's likely, though, that after two or four bins, you will say "enough!" Few people get the worm-crazies like me and I'm sure on some level you or your spouse will be thankful for that small mercy. So what should you do whenever you get to the point of having too many worms?

What you don't want to do is just put the "excess" in your same old bin. The general rule of thumb is always going to be about a pound of worms per cubic foot of bedding to start with, regardless of if you have a small indoor bin or a larger outside windrow. Commercial operations will run a little differently and that is a whole other book; I'm only going to talk about you growing worms for yourself and maybe some extras to sell occasionally, not a big operation. You can fudge on this by about a half pound, but too many more than that and they will get overcrowded fast. "Overcrowded" will depend on the size of your bin because again, you won't really know how many worms you have except at the beginning and at the harvesting of every bin. A good time to discern if you are at or near the point of being overcrowded is when you finish your weekly mixing and maintenance. If, after everything is all mixed up, you reach in and pull out a handful of whatever from anywhere in your bin and you have more than about 30 or so worms in your handful, then you are at or quickly approaching overcrowded conditions. So putting your "excess" in with some other worms is not the best solution. At some point, if you don't take care of it first, your worms will let you know you have exceeded maximum density because they will start trying to leave to find less crowded conditions, or die trying.

Instead, if you don't want to start a new bin for the extras, either give them away or sell them. Which you choose depends on many things and is a personal choice. Craig's List is a great place to sell them. Vermicomposters.com has a function on their member map which allows you to flag your location if you have worms to share or sell. Freecycle groups are another great place to find someone in need. Sharing worms with other beginning wormers is a great thing in my book. I'd love it if people giving away extra worms were as commonplace on Craig's List as people getting rid of old couches. That would be so cool! It is also possible that your original supplier would be interested in buying them back. I buy back, but so far everyone who has bought my worms has decided to keep them all for themselves or give them to friends. Vermicomposting is a little contagious and there isn't a single thing wrong with that. As such, the very best thing to do with your extra worms is to get one of your friends to start their own worm bin and you can give them their "seed stock" and a copy of this book to get them started.

Worm Sex

(Boom Chicka Whaa Whaa)

I'd so love to tell you that this is where things will start getting spicy and that if you are under 18 you should probably talk to your parents or guardian before reading this part...But I can't. It's *worm* sex, people. "Scintillating" is not the first word that should come to mind. Sorry, this is not that kind of book. It's kind of interesting though. That will have to be good enough.

You *want* your worms to have sex. As much as possible. More sex = more worms. Worms having lots of sex is also a great indicator that you are doing things right and your worms are very, very happy. Though I will tell you, sadly, that you may never actually *see* your worms getting, um, intimate (gotta be a little proper, there are under-18 folks who are going to read this book after all-but I promise I'll only be a *little* proper).

You may have noticed the dearth of illustrations in this book. I was going to have photos, but then I realized that a photo of wet paper shreds or of worm poop itself is not really so helpful and not worth the extra money you would have to pay for this book if I did do that. Then I thought I'd just draw the illustrations. But since my drawing a worm bin or a pile of food would essentially be the same thing because I'm not so skilled in that area, I thought I'd just make sure I'd describe everything well enough so that you, dear reader, would get a good visual image of what the heck I'm talking about for any given situation. Someday I'll have a website and have pictures on it you can go see. If I *did* have pictures, now would be the point in the book where I would have a handy illustration of a worm and it would have little areas pointing to different parts of the anatomy so you could become that much more familiar with the worm itself.

Then I decided: booooring. I've read tons of worm books and articles and all I have ever done is glance over those illustrations. And I LOVE worms. If you are very disappointed right now and really, really wanted to see a worm anatomy illustration, put the book down, go to the internet and search "worm anatomy" and feast your eyes on a zillion such illustrations provided by far more skilled artists than I will ever be.

For the rest of you, I will briefly describe the anatomy of an earthworm. I do this because some of the worm parts are actually relevant to this discussion. And I'll be brief so I don't lose you altogether.

Get a picture of an earthworm in your head. Got it? Okay, see that both ends are practically identical, but one end has either a band of different colored skin nearer that end or a band of different colored skin that is raised up and makes a road bump? This band, regardless of what it looks like, is called a clitellum. This is where two earthworms connect after a couple of drinks in a dark, smoky bar to make more worms. Okay, maybe not the drinking and smoky part, but definitely the dark part.

There is a lot of folklore surrounding how worms reproduce. I can't imagine for the life of me why this would be, but let's set the record straight once and for all, shall we? First of all, worms are hermaphrodites, meaning they have both ovaries and testes for sexual reproduction. These reproductive organs are located at the end of the worm nearest the clitellum. This does not mean, however, that they mate with themselves. I did read in a hoity toity soil science journal once that some varieties of worms actually *can* do this in very extreme conditions (like: they are gonna die and can't find a mate), but that is *some* worms in *extreme* conditions (though frankly I'm not even sure how the heck they could possibly know this), it certainly doesn't apply to your worms. Your worms will only be able to reproduce with another worm. And that doesn't mean just any old worm, but only a worm of its own kind or of a closely related kind (the evidence for the "closely related" part being somewhat in dispute, mind you). This means your night crawlers are not going to mate with your wigglers. Only crawlers with crawlers and wigglers with wigglers. Some crawlers with others and so on, but considering how many different kinds of earthworms there are and how hard it is for us to discern most from each other in the first place, I'll just leave any further discussion of who the worms will and won't have sex with to the aforementioned hoity toity journals.

Another myth about worm reproduction is that if you cut them in half, both halves will grow into two new worms. Poppycock! I don't even know where someone would get this idea in the first place. I do know that you can cut a worm in two and the part with the clitellum, depending on how much of that is left, will still live. But the poor little amputated worm will not be doing much of anything else (read: making poop or making more worms for you) until it is all healed. How long that takes is anyone's guess and will definitely include the phrase, "it depends". So NO CUTTING YOUR WORMS IN HALF. You worm sadist.

Okay, back to real worm sex...When two worms find their "Mr/s. Right Now", they each develop a thick mucous around their clitellum. They then sidle up to each other, going in opposite directions and attach to each others mucousy clitellums. (Just stop it right now. This isn't easy to *write* with a straight face either.) At this point they *both* exchange sperm with each other to fertilize each others eggs.

As I said before, it is possible that you will never see this happening. Worms like their privacy, just like you do. But I'll bet you five bucks that if you ever actually *do* get to witness this exchange, your first reaction will be that you are witnessing a worm in the process of dying. (Perhaps not so different from us after all, eh?) Everything I've *read* says that the worms sort of scootch up next to each other and then connect. But having been privileged to have actually seen this several times, I can tell you this is not exactly right. They never stay together long enough once I try to get a good , up-close look (would you?) so I can't exactly say what they are doing, but a mere scootch it is not; at least with any of the several varieties I have at my farm. What they look like, when you finally get to see it, is that they actually are tied in a knot at the clitellums (hmmm, clitellii? Heehee). Sort of stuck like each one is entwined under the band of the other. Once you get a decent look at your "dying" worm, it will seem more like the two worms are trying to make a bow tie of themselves. At first, it will look like one long worm with two fat tumors growing out of its middle. Then, they will realize you are watching (or more likely their photophobia kicks in, and by that I don't mean they are afraid you are filming them) and they slip apart and you, like me, will have to wait and wait for another good opportunity to finally figure it out. Lemme know when you do, k? You can actually google an image of "earthworms mating" if you want to see it. You'll see the side-by-side type photos but also a couple of the "bow tie sex" photos. But, um, please don't google "bow tie sex", okay? I don't think you will get worm pictures if you do.

Lest I inadvertently start a whole new chapter in the annals of worm folklore, by likening the process to looking like the worms are dying I am merely giving you an indication of what you will be thinking when you first see it. I am not, by any means, saying or implying that your worms will only mate once and then die, okay?

Between the time I wrote the above and the time you are reading this, I did google this image. (And I was amazed that the auto-finish feature kicked in when I was typing "earthworms mating"; apparently I'm not the only one looking for images of worm sex. Frankly, I'm not sure whether I should be happy there are so many kindred souls or very, very afraid.) In any case, one of the images perfectly illustrates what I just described, but another does show two earthworms just scootching up to each other like I've read in other places. The scootching worms look nothing like mine and their clitellums are much closer to the middle of their bodies, too.

That just goes to show ya that "earthworm behavior" really does "depend on" like I keep saying. Apparently the kind of worms you will have will determine what, if any, mating behavior you do see.

And I'll add that the "if any" part must definitely be true for people besides me since it is the middle of 2010 as I write this and there were maybe four actual photographs on the whole of the internet that showed two earthworms mating. There were several pages, but most were drawn illustrations and the majority of the rest were the exact same picture over and over again from different sources. Which really goes to show you how "knowledge" is spread in the internet age. There was only one photo of the kind of mating that I have seen, the rest were of the scootching kind of mating. Everyone referencing the scootching image when they write about worm sex are spreading the idea that this is the way that "earthworms" (read: all of 'em) mate. But now at least you and I will know there are at least two ways. Maybe more. (It's also interesting to note that one of the articles associated with one of the pictures I saw dealt with a study involving the measurement of the amount of sperm deposited when the worms mate. The study mentioned that their mating can take *up to an hour*. Look, I'm just the messenger. Who knew? My earnest apologies to all the worms that experienced coitus interruptus because of my voyeurism. Mea culpa to be sure. And if you are under 18 and don't know what that means, please don't go ask your parents or they will worry about what kind of book you are reading!)

Back to the original subject...Once they exchange their sperm, the thick mucous coating around the clitellum will begin to harden. As this happens, it will be pushed down to the nearest end of the worm. The ovaries release the eggs and both mix up inside the mucous ring, which by then is getting harder on the outside. The whole shebang slips off the end and at that point it is a worm cocoon.

Your worms are quite promiscuous and will mate quite frequently with whoever strikes their fancy. How often do they mate? I dunno. I'm not sure really that anyone does for sure or can really say with any certainty. Counting your worms at the start and finish and doing some math can give you and average of how often they *successfully* mate in the conditions of that particular environment. Or, perhaps more accurately, how often successful pairings result in successful progeny...Point being, you will read all sorts of things about this in other places, but it will always come down to how well things are going with what *you* are doing for them in *your* bin. Some of you will be wildly successful and blow all previous estimates out of the water (you're the CHAMP!). Some of you will not be quite that successful. The rest of you will fall somewhere in the middle. Why? Because that's what the middle is for: most of us. Fret not, fellow wormer. You will be happy with the results regardless because the poop will work well even if there are 8.46% fewer cocoons in your poop than in the poop of the guy down the street. Promise.

The cocoons themselves are actually really cool. They are hard to see at first. When I saw my first one, I thought it was a seed from something I had put in as food and I was wracking my brain to try to identify it. Silly me. Lest you fall into the same time wasting brain teaser, I'll give you a heads up so you can marvel appropriately when you see them.

The cocoons, also known as "eggs" by those of us who get tired of writing "cocoon" all the time, can be any color in the range from light yellow to dark brown. As the egg matures, it will turn from light yellow to gold to tan to dark brown. They are mostly round and super tiny, only a couple three millimeters long by a couple millimeters wide. They are not entirely spherical. One eensy tiny end is pointed a little bit. I think this is because of how they slip off the pointy end of the worm, but that is just a guess. Wiggler eggs are about half the size of night crawler eggs. If you have only or mostly night crawlers, it will be much easier to see the eggs when you are doing your bin maintenance. If you have only or mostly wigglers, you will have to train your eyes a little bit because the eggs are so very small. Once you see them for the first time, though, they will be much easier to spot in the future. Sometimes you will see cocoons mixed in with the worms when you first get them. More likely, you won't see them until your bin is about half turned to poop or more. This isn't because they aren't there, they are just much easier to see against the dark background of worm poo instead of the light background of shredded paper. You can also find images of worm eggs on the 'net, if you want a better idea of what to look for when you get to that point.

How many baby worms are inside each cocoon is also a matter of which source you want to believe. Some say one, some two, some four (I've never read just three, isn't that funny?), some six or seven. Make up your minds. Argh! My logical guess would be the number is closer to two, but I'm quite sure that depends at least somewhat on the kind of worm. It may be that there really are seven in that tiny little thing, but only one or two are ever expected to survive. I'm not sure we'll ever really know. I am sure that it doesn't matter. Worms mate a *lot* in good conditions so one baby or seven, if you are doing it right they will be too and in turn you will get poop and more worms and that is what really counts.

And yes, even though they only mate with their own kind, this will all still happen and happen often even if you have a bunch of different types of worms in your bin. They will still meet the ones of their own kind and just pass the others by with a fine "how d' ya do?". It is possible that there may be a slight decrease in the percentage of new worms overall compared to a bin of all one type of worm. But the trade off (having diversity to overcome any potential problems that will only affect one type of worm) far outweighs any potentially slower reproduction rate. I've never noticed that there were fewer worms than I expected when I harvest a bin with multiple varieties, though, and I doubt you will either.

One thing that you can do to facilitate more romantic encounters in your bin, besides just keeping the conditions livable, is to give the worms what I like to call "love shacks". Yes, worms live in a dark, moist environment. But apparently to them there is dark and moist and then there is *dark and moist* (sorry, I couldn't find a "romantic" font so italics will have to do). I've found more worms making whoopee in the love shacks than just in the general population, so while my words might be tongue-in-cheek, it's serious business for the worms.

Love shacks are areas that are dark and moist but are *not* poop or bedding. One of the easiest shacks you can make are stacks of egg shells. So many people give me the shells already stacked in four or more half shells nestled into each other, so all I have to do is put them in as is instead of crush them. I only do this when there is plenty of other grit in the bin and these shells would just be "extra". Other great love shacks are ears of corn. Worms LOVE corn cobs, especially when there are still little bits of corn left on the cob. I bury the cob in the bottom and after about a week it gets soft and I break it into two or three shacks and bury those. The worms eat out the inside of the cob, the corn "marrow" if you will, and that leaves an air tube in the middle of the cob. The air tube is the worm equivalent of an hourly rate motel. Pineapple tops are a big hit, too. They hang out near the center and get to know each other very well until the top becomes soft enough to be food, then they eat it. Stacked avocado skins are another good one and they will last a very long time. They really love to be between moist pieces of cardboard, too. And also in the little corrugated tunnels, but these only work if the worms meet and are facing the right directions (wink wink). I think that is enough to give you a great idea of what makes a good aphrodisiac. Love shacks don't take any extra effort, but are definitely worth having.

Feeding in the center is not only good for you to be able to monitor the worms food intake, but feeding in the center (instead of all over one side, for instance) gives the worms a central location to meet. Sure, lots will be in the corners (no, I don't know why they like to hang out in the corners any more than I know why teenagers like to hang out on corners) and quite a few will be in the general bedding. But many, many will be hanging around the "kitchen" and having plenty to eat and also plenty of opportunities to meet other worms who are both ready and willing.

I do know that in order for a worm to actually make these wonderful little egg packages for you, they have to be mature enough to have sex in the first place. Remember at the beginning of this section when I talked about where the clitellum is and what it looks like? If your worms don't have that clitellum, they aren't old enough to make babies. This can, however, be a little tricky since the clitellum on some types of worms is not so easy to see. On some, it makes a bump and is readily visible. Sometimes the bump itself is a whole other color from the rest of the worm. In others, the clitellum is only expressed as a slight change in skin color with no bump. Others have an obvious change in skin color, but still no bump. Younger worms of any kind, however, will look pretty much the same from stem to stern, so if you are looking at a worm and wondering if it is old enough, the clitellum is what you will want to see.

A lot of people want to know exactly how long a worm will live. I do hate so much to sound like a broken record but...Like so many other things, reports in both popular and higher literature disagree. The answer, if there is only one, will also depend on their living conditions. I've read anywhere from several months to a couple of years. Again, I can't for the life of me figure out how anyone would know. It is not like there is any kind of "catch, tag and release" program for worms in the wild. No mini-radar-collar one can attach to gather data with. Same would hold true for the worms living in your bin. No, sorry, you can't paint one with a little bit of nail polish and track how long it lives. Won't work. I can tell you that I have a lot of worms and they keep making me more and more and as long as they keep doing that, I'm happy and I'm guessing you will be, too.

Bugs and Others in My Bin

(or, What the Heck is THAT?)

Okay, squeamish people. You've made it this far, you can make it through this part, too. Will it help to tell you that this whole section is all part of a perfectly natural process? No? How about: none of these bugs will actually hurt you. How's that? Better? No? What if I said that most of these non-worm life forms will never leave the confines of the bin? Sorry, I probably put you in an anxiety attack by saying "most", didn't I? Look, I'm not sure what to tell you other than it is unlikely that many, if any, of the larger non-worm life forms I'm about to talk about will ever even *be* in your indoor bin. Most will likely be not much larger than a comma on this page. (This big: , See, that's nuthin'.) Okay? Have no worries, you'll buzz through this section none the worse for wear.

When people talk about worms eating garbage, they technically aren't talking just about the worms. There are some other guests that help in this process and some that may not help much but at least are not harmful. Most are completely harmless. A small few are just pests to humans and sometimes even downright annoying. Annoying happens every day, but in this case I'm going to help you figure out how to deal with this particular type of annoying so you can live in perfect peace and harmony with your worms. (Can I get a "kumbuya" please?)

Your worm bin is not merely just a place to raise worms to get their poop, it is actually an entire little ecosystem in a box. There will be millions of microorganisms in your bin and even some not-so-micro organisms. All of these work *together* to give you the gorgeous end product we call worm poop. But it isn't just worm poop, it's the poop of a bunch of other things, too. Your plants need all of it, so try to bear with me here and not freak out. I think when you get to the end of the book, you will be much happier with this whole idea than you may have been at the first moment you read "bugs". It is a far more fascinating process than just a little worm home, let's dig in and learn more about our other composting buddies...

Bacteria

I've already mentioned different types of bacteria several times. In this sani-wiped, anti-bacterial-sprayed, germaphobe world we are now living in, the mere mention of the word "bacteria" causes some people to get the heebies. But the problem with these heebies is that sometimes bacteria can be bad and sometimes they can be good (sometimes even great). The anti-bacterial crowd isn't known as the "anti-bad-bacteria" crowd and that is a shame because we need some kinds of bacteria for good health and so does the soil and so do plants.

Bacteria, be they the good kind that come from when you are doing things right, or the bad kind from when your bin conditions are screwed up, will always be in your bin. The "bad" kind generally aren't bad because they are going to be harmful to you, only to your sense of smell (unless you decide to eat them, but trust me on the smell thing, you won't be tempted). Anaerobic conditions, like when you get a bunch of water in the bottom of your bin because you overwatered and didn't correct that situation quickly enough, is the most common source of the stinky bad bacteria. "Air" is the cure for the bad bacteria in your bin, NOT anti-bacterial cleaners.

The good bacteria population will rise and wane and the types will change somewhat as the conditions in your bin change. Some of them will help break down the food that you put in the bin and make it easier for your worms to eat. Some of them will get eaten by the worms. Some will get pooped out by the worms. All of them will work together to make great poop for you to use and they will become a part of what makes worm poop so valuable to your plants. They won't hurt you one bit and you'll never see them, so let's not worry about them one bit either.

Mites, Springtails and Potworms

Some things, though, you will see. But they will be tiny, like the comma I just typed. Or the period I typed just after that and am about to type again. (Just in case you weren't sure how big that is.) Some might be *twice* that size. A few are even as big as this "o". Oh my. See, squeamish people, there is nothing to worry about. They are tiny *and* harmless.

The small bugs you will most commonly find in your indoor bin are mites, springtails and potworms. Potworms don't really fall into the little bug category because they are actually worms and worms aren't bugs. Mites, springtails and potworms are all totally natural in your bin and rarely pose any kind of problem that needs addressing.

Let's start with springtails. Springtails are itsy bitsy little white or near white bugs that sometimes have a "tail" that actually allows them to jump a wee bit, hence the name. But not all of them jump and even the ones that do don't jump much. They are super tiny after all (and everybody knows white bugs can't jump! haha). At times, they can be more prolific and visible than at other times, depending on what you are feeding and the moisture conditions of your bin. They eat decomposed stuff and some molds just like your worms do and help make the vermicompost yummy for your plants. There is zero reason to worry about these tiny creatures whatsoever. Besides, there isn't anything you can do to "get rid of" them other than starve them and starving them would starve your worms so fuggetaboutit!

Mites are another tiny creature that will naturally be present in your bin. They range in color from very light brown to reddish brown and even some can be brighter red with no brown tinge. They are super tiny and round and move so very, very, very slowly that you will have to concentrate to see them do it. These are NOT the mites that might bite you and make you scratch until your skin is raw. Mites in my bin tend to congregate on carbohydrates like bread and tortillas and the like. They like cardboard, too. Generally they don't congregate forever and you may have them all over your bin and you will never ever see them unless they are lounging on a big piece of bagel. They also contribute to the decomposition process happening in your bin and are a totally natural worm bin guest.

I've read all sorts of crazy internet posts about mites in a bin. I'm not sure where this rumor started, but: enough already crazy worm people! Mites are part of the process and help break down and eat some of the decomposing matter and similar detritus you throw in your bin. They are not going to hurt you or your worms. These are not predatory critters. Sometimes you see mites and/or springtails all over a newly dead worm (You will know it is newly dead cause if it had been dead for too long it would be worm soup and you would never see it.) Just because these teensy bugs are on a dead worm does not mean they killed it, okay? Maybe they just walked in and the worm was already dead and they just picked up the gun! Seriously, they are on the worm because they eat dead stuff. The worm is dead. One dead worm totally happens. Sometimes they are old or maybe they were never right to begin with, it doesn't matter as long as you don't see more than one or two and never very often. You won't see many dead worms since you will be treating them right. But when you do and when you see mites or springtails on the worm, rest assured the worm died of something else, not some 50's era "Attack of the Killer Worm Bin Creatures" type scenario.

Potworms also eat decomposed stuff. That's why they like to hang out with your worms. Potworms are about a centimeter long and white or creamy white in color. They will hang out all over the place, but mostly will hang out right where the food is. Their population is usually higher in wetter bins. Potworms are completely innocuous. They won't hurt you, they won't hurt your worms. Sometimes it will seem like there are TONS of potworms in your bin. If so, you likely have either too much food or too much water or both in your bin. You already know how to fix that so you can view a potworm population explosion as a little bit of a barometer for the conditions in your bin.

I get calls sometimes from people who tell me that there is some kind of problem with their bin. They tell me that they can't seem to find any worms, but that they can't have done something wrong because "there are tons of baby worms in the bin, just no adults". When I ask them what color the babies are, they tell me they are white. There are no white baby composting worms. Baby composting worms (earthworms) are pink or red. White "baby composting worms" are really potworms, which are much easier to care for than compost worms, but don't give you the same kind of poop that compost worms do which is why we don't do potworm composting.

Potworms are also not nematodes. Nematodes will be in your bin, just as they would naturally be in your soil. These kinds of nematodes are going to be the *beneficial* kind and are also microscopic, so you will never know they are there except in principle.

People frequently wonder where in the world the mites and potworms and springtails come from to begin with. After all, all they put in their bin was shredded paper and worms, right? Well, no. Squeamish people warning: you may not like this on the surface, but thinking the following *through* may actually cure you of your squeamishness...Read what follows with that in mind, okay? If you like being squeamish, skip the rest of this paragraph. Potworms, mites and springtails get in your bin the same way fruit flies get in your house: their eggs are on, and even sometimes in, your food. Did you just "ewww" out loud? Oh, stop it. You have been eating every day for however long it took you to get this old and your food has always had itsy bitsy things on it you can't see and that don't hurt you one whit. Some of those things will grow into other things in the right conditions. Lest I inadvertently start yet another urban legend, let me be very clear on this point: springtails, mites and potworm eggs on your food will NOT grow in your body. They *will* grow in your worm bin or compost pile and the like because conditions in those places *are* right for them. So stop being such a pansy about it or I will tell you where gelatin really comes from.

Flying Bugs

There are other bugs that you will be more than happy to be upset about and I'll totally support you. Flies. Argh! Flies are the bane of my existence. I am surrounded by livestock poop and rotting food so flies are definitely on my mind in the warmer months. No, silly, flies will not automatically be a problem for you just because you have worms. Just sometimes and only for some people, not always.

Fruit flies best NOT be a problem for you because you will be freezing your food before you give it to your worms. Other flying things, though, can be pests. Flies will be a pest if they can get to your bin. If your bin has a fly-tight cover (make sure it can still allow air for your worms!) then this will likely not be a huge concern for you. Just another reason to take your bin construction under careful consideration. If you also keep your worm bin in your house and you traditionally don't have a huge fly problem in your house, they will likely not be a big problem for you. If you keep your worms in the garage or some other place where flies seem to congregate, then this can be part of the problem.

Flies love rotten food. Their kids do, too. It's not such a huge problem that the flies might get to the food and eat some. The problem is that the flies recognize how super cool the buffet is that you left for them and know that is also a wickedly perfect place for their kids to be born and raised. Fly kids are maggots, just so we're clear here. Blech to the nth degree to be sure. As such, you want to make very sure that you do what you can to prevent any problems before they start lest you be completely freaked out the first time you see some "rice" in your bin and can't remember when you last had leftover rice. (Yes, I know: gag.)

If you do get flies, there are a few relatively easy steps you can take to alleviate the problem. First, make sure that your bin, bin lid and the location of your bin are not contributing to the fly issue. The bin should be constructed in a way that does not allow the flies to enter the bin. Generally any weak point in the construction will be in the air holes or whatever "system" you designed to allow the worms to get enough air. Make sure any screening is completely flush with the lid. You would be amazed what tiny holes the flies can get through.

This, right here, is the #1 reason that you did not want to put holes in the sides of your bin. If you constructed your bin the way I recommend, the lid will be the only point of entry and it can easily be made fly tight with some adhesive weather stripping around the edge of the bin. Also make sure there are enough staples holding the screen or landscape fabric tight onto the lid. The only other problem could be if you used screen with holes too large. If you already made a bin with holes in the side before you started reading this book, you do have a few options, though the best option is to start over with a new bin if none of the "fixes" work for well enough.

The first thing is to, obviously, cover the holes in the bin. Before you do this, though, make sure your lid arrangement will still allow for air access. Clean the area around the holes on the outside of the bin with some rubbing alcohol and try to see if duct tape will stick over them. If so, you're golden. But tape doesn't always stick well to plastic so this may not work for the bin you have.

If not, the next best bet is to try to staple some screen or fabric over the holes. The problem with this is that it is nearly impossible to get the staples securely into the plastic if you are doing this from the outside because you do need a stable surface to staple into. Which means that, in order to do this properly, you will first have to completely empty the bin so you can lay it on its side. This is the only way to have something stable on the opposite side from where you are stapling. If you do this, make sure if you are using metal screen that you fold over the edges once or twice before you staple the screen on, otherwise you will cut your hands on the sharp edges when you are doing your weekly mixing and maintenance.

If that doesn't work, there aren't many other things you can do that are not just going to be variations on this same theme. No, sorry, you can't just stuff the holes with some cotton balls and call it good. Bugs, especially flies, can totally still get in. I suggest getting a new bin. One without holes in the sides.

If your bin is as fly-tight as possible, make sure that you are not leaving your worm food out somewhere that a fly or two can get to it and lay eggs on the food before it goes in the bin. Doing so allows their kids to be hatched right in the bin even if the adults can't get in there. A nasty surprise if you are sensitive to that sort of thing. This is also yet another reason to keep your compost can in the freezer. If for some reason you find live flies in your freezer, you will want to talk to your local entomologist, not me. Lastly, make sure you are burying your food when you give it to the worms. Flies don't burrow, but their babies will if the eggs are laid on top of your bedding. So again, it's best to make your bin as fly-tight as you can from the git go.

If for some reason you correct these problems and still have a fly issue, it is time for the defense to take a rest and let the offense take over. One easy fix is to just pick out the baby flies when you see them in your bin. They are pretty easy to find and generally congregate all in the same area. I don't recommend trying to smoosh them, they are tough little buggers and can make the kind of mess when you squeeze them that will turn you off of worm composting forever and ever in no time. Instead, just put them in a cup or container and toss them out in the garbage or put them in the freezer for a half an hour. Or feed them to your chickens. Problem solved.

If that idea is just too much for you, traps are very effective. I never recommend using any of the smell attractant fly baits. None of these are designed to be used indoors and can literally kill small animals and make large animals very ill.

Old-fashioned fly paper works wonderfully and is very inexpensive. The only thing is that you do have to remember to change the paper every couple of weeks or it tends to lose its "stick" (unless you walk into a low hanging one and it gets in your hair...not that this has ever happened to me, of course). They make a new kind of fly paper now that is much less obtrusive. It is just a clear, flat sheet that you can stick just about anywhere. I've even stuck these right to the underside of my bin lids and they work like a charm. You can get four in a pack for less than two bucks and they never get stuck in your hair.

You can also just let nature take its course. For most folks this will not be enough, but if you have a lot of worm bins and uncovered garbage and manure laying about, you find that the flies actually leave you alone for the most part since there are so many other, more tasty treats nearby. Believe it or not, the poop from the baby flies is just as good for your plants and also helps to make naturally good soil. I'm told, but have never confirmed, that there are maggot farms in certain parts of the world just like there are worm farms. Some things I don't even *want* to research! If you ever visit one, I'm happy to see the pictures, but please no olfactory descriptions.

I have read some things that say you can stop feeding your worms for a week or two to help starve the maggots. I have personally not found this method to be successful since there are always going to be little bits of food in your bin from the mixing. These little bits are more than enough to keep the little buggers alive, unfortunately. You can try it and not starve your worms since they should have bedding and the little bits to eat, but if you find that it doesn't work, please don't expect me not to say, "I told ya so!"

There are also any number of homemade fly traps you can find directions for on the internet. Some work, some not so much. But since each trap has its proponents and detractors, I will leave the decision of which homemade trap you want to try up to you.

One of the most effective ways to go on the offensive against too many flies is to use the flies' natural predators against them. The term is "integrated pest management"

(IPM) and what it means is using nature's own resources to your own ends. But in a good way, of course. Fly predators are available by mail order from an insectary or in person during fly season from a number of livestock supply stores. An insectary is a business that raises beneficial insects to help combat infestations of bad bugs. (And you thought raising worms was a weird business to be in.) There are tons of beneficial insects available to help you fight any number of bad bug issues. If this is an area you are unfamiliar with, please take the time to do a little research to find out if a beneficial insect will work for a problem you are having in your yard or garden before you rush out to buy some laboratory-invented chemical to take care of it instead.

Having flies is the worst thing for some people. Fortunately they are rarely a problem requiring too much thought unless you get to a larger scale like me. Seriously, I know I covered this pretty well, but only "just in case" for you, it is not likely that you are ever going to have a fly problem at all, much less on a scale like they are for me. They are really only a problem for me because I usually have about thirty different kinds of bins going at any one time and I keep them in a garage-like breezeway nine months of the year. I have this many types of bins for teaching purposes, to show what works and what doesn't. Hence, I have a fly problem. But only for educational purposes, I swear. If you end up with this many worms to take care of and don't want to use an outdoor windrow or the like, you can preplan your system to take better care of the lids and such to avoid the fly issue altogether. I have zero fly problems with my windrows, so don't worry too much if that is the route you eventually want to go.

Fungus Gnats

Flies are actually not the thing that drives me nuts the most. Fungus gnats are. Fungus gnats are like itty-bitty fly wanna-be's. They are black and tiny and you will know they are not fruit flies because they generally don't hang out by your compost can and because fruit flies are brown and three times as big. Fungus gnats are totally not harmful to you. Just absolutely annoying, especially when one lands in your beer. Argh! I just hate that.

Fungus gnats feed on fungus. Pretty simple. Fungi will be naturally present in your worm bin as part of the decomposition process. Molds, too. But not the kinds of molds and fungi that cause people to move out of their house. These are the good kind that you breathe in and smile over when you get a whiff of really good dirt. Fungus gnats smile at that, too. The good news is that they are relatively easy to get rid of. But *only* if you do something about it at the *very* first sign of a problem.

Fungus gnats like the decomposing matter, but only if the moisture is right for them. Their life cycle is very short and they have to have some nice dampness to thrive and for their kids to be born and thrive. Fungus gnats generally mean your bin is too moist. At the first sign of fungus gnats, add some dry bedding to your bin and/or add a two inch layer of dry paper to the top and pat it down to make it a true two inches (instead of just one inch fluffed up to two inches). Next, go check all of your house plants. Fungus gnats can start there, too, so make sure your houseplants are not overwatered. If they are, dump or soak up any water in the catch tray. If possible, remove at least the top two inches of soil in your houseplants and replace it with new. Lastly, stop overwatering your houseplants.

If all of this doesn't alleviate the fungus gnat problem within a week at the most, it may be time for a trap or two. Keep up with the drying process though, as that is the easiest way to get rid of them. To trap them, you have two common options. One, go after any of the adults you see with the hose of your vacuum cleaner. Suck 'em up then empty the bag outside or stick it in the freezer for twenty minutes. Keep this up every day (their life cycle is super short) until the problem goes away. You can also use the aforementioned fly traps, either kind. The new, clear ones are of limited value though. If you find these and prefer them, get your hands on some bright yellow paper, too. Cut a piece of paper the size of the flat, sticky trap and tape up the yellow paper behind it. Fungus gnats dig yellow and will go there first.

You can buy sticky traps already made out of yellow paper at most garden centers. But they cost almost three times as much and the only difference between these and regular fly traps is the yellow backing. You can get a whole package of construction paper at the dollar store and there will be more than enough yellow paper in there, or just draw yourself a sheet of yellow in your photo editing program and print it up. Voila! I've also used honey smeared onto a sheet of yellow paper, but I don't think this works quite as well and it is a total cleaning drag if you use too much honey and it drips off the sheet and onto your carpet.

If for some reason none of these things work, then it is time to bring out the big guns and order some predators from the insectary. It is highly unlikely that you will have to take this step unless you have a LOT of indoor bins. But if you do, then IPM is definitely the way to go.

Most people will never have a problem with fungus gnats unless you add leaves and/or pre-decomposed compost or manures to your indoor worm bin. There are lots of fungus gnats outside, but because outside is so much more vast than your home, you likely never notice them. Outside, they tend to lay their eggs in places that are very similar to your worm bin, like your compost heap or a bit of leaf mould (decomposing leaves). If you find that fungus gnats are a recurring problem for you, stop adding leaves and compost and stick with things like shredded paper and food only from your kitchen and not things like rotting crab apples from the ground outside.

Speaking of fungus...It's likely that occasionally you will find some in your bin. Either in the form of a bit of visible mold or in the form of actual mushrooms. As I mentioned before, your worm bin is a whole little ecosystem and molds and fungi are a natural part of that system. These are not the kinds of molds that cause you to get tremendously ill, more like the kinds of mold you will find if you leave some leftovers in the back of the fridge for too long. These are also *not* the kinds of mushrooms you want on your pizza. Generally these molds and fungi will sprout up when your bin is both slightly too moist and when there are a lot of carbohydrates in your food supply. Just fold them back into the rest when you do your mixing and keep your bin a little less moist and this "problem" will quickly go away.

I've had problems with vinegar flies only once. These are a lot like fruit flies in size, but look more like really small regular flies. The same preventative measures for all of the above pests should be addressed first. Then, to catch the grown ups, put about two inches of cider or red wine vinegar in the bottom of a glass or jar. Add a couple three drops of liquid soap and swish it around to mix it. They'll practically swarm this treat, but will drown in the vinegar since you added the soap. The soap breaks the surface tension so the little buggers can't just sit there and suck up the yummy vinegar. You can add more vinegar as needed, no need to empty the whole thing until they are all dead or you get sick of looking at it and want a fresh jar. This vinegar and soap bath works somewhat for fungus gnats, too, but doesn't get them all.

Well, that's it for the flying bugs. There are a few more kinds of other bugs to think about, but these last will rarely ever be in your indoor bin and are generally confined to outdoor endeavors or to bins with "outside" stuff used as food or bedding.

And Others

Centipedes and millipedes are very common in an outdoor worm bin or compost heap. Millipedes, which have two pairs of legs for each body segment (sorry, they don't really have a thousand legs!), are not a bother to anyone or anything except maybe some delicate sensibility you have. I have read that centipedes (which only have one pair of legs per body segment) can sometimes eat worms and their eggs. While true, the good news is that centipedes are loners and very territorial. If you find one, kill it. Chances are very good that there will not be another one around the immediate vicinity so no need to worry about an "infestation". Just keep your eyes peeled for the next usurper who will try to move into that territory.

Pill bugs, sow bugs, and roly-poly bugs are all just about the same bug and all completely harmless. They will be in with your worms if you use manure or leaves or other "normally outside" items as bedding or food. They are a natural part of the process and won't hurt a thing so just enjoy watching your toddlers' eyes widen when they see them roll up in a ball and leave them alone.

Grubs are by far the ugliest of all critters you will find with your worms. If you use livestock manures (the big animal manures, not poultry or rabbit manure) to feed your worms or as bedding, you will have grubs. Grubs are just baby beetles and come in a TON of varieties. They generally are hugely fat, caterpillar-like blobs that range from white to cream to gray in color. They almost always curl up in a ball like a roly-poly if you disturb them, but they are much larger and not scaly like a roly-poly is. Grub poop is good stuff. The problem is that some beetles are dangerous to area crops and some are not. I haven't been able to find an easy way to tell them apart at this stage. Most grubs look exactly alike. If you find you have a lot of grubs, it would be a good idea to talk to your local extension agent and/or your local farmers to find out if you should be exterminating them or leaving them alone.

Depending on the conditions where you live and on where you keep your worms, you may find that your bin is attracting ants. Ants like the food you are feeding the worms, not the worms themselves. Don't freak out, it is unlikely you will have ants in an indoor bin unless they are otherwise a problem in your house. There are a zillion natural methods of getting rid of ants; a quick internet search will give you more options than you need. Setting your bin on blocks and then putting the blocks into little tubs of water works well, as does setting the bin on blocks and then smearing the blocks with petroleum jelly. If the ants can't get to the food, they'll find somewhere else to go and leave your bin alone.

Lastly, you may find that some types of spiders are attracted to your worm bin or windrow. If I were a lady spider, I'd find a worm home to lay my eggs, too. It's safe, it's dark, what more could I want? If you have spiders in your bin or windrow, you may not even see them. But sometimes you will see the baby spiders just after they hatch and you may very well freak out a little bit (okay, likely a LOT) if you have never witnessed this before. It will seem like there are MILLIONS of those little suckers in there. In reality, if you can make yourself think rationally when that happens, there are only a couple hundred and most of these will not live out the week. Trust me, the vast, vast majority of them will die a natural death and the few that remain will help keep down the overall bug population in your yard and garden and, by extension, your house. I've only ever had spiders in bins that are in my breezeway or in my basement. Never when I've had a bin in my kitchen or the like so please don't go all arachnophobic on me.

Still with me? Awesome! You, grasshopper, have shown you have the fortitude and willingness to go forward with this wonderful worm adventure and as such, reap the tremendous benefits befitting one of your cunning and grace! Read on...

Section Three:

The Poop, the Whole Poop

And Nothing But

-harvest time

-reaping the harvest

-poop 101: basic to advanced poop use

-sweet, sweet poop tea

POOP!

Wow! I'm so happy you made it this far. This is what it is all about. The poop, literally and figuratively. I can hardly say enough good things about worm poop! (Yes, yes, of course I would say that.) I think I've already covered this, but just in case you are one of those people who skip around a book instead of reading the whole thing cover to cover (a shame) I'll cover it again...

Worm poop is a naturally occurring resource in healthy soil. Some feel it is the soul of the soil and I'm not one to disagree with that too much. Healthy soil is biologically active and contains far more than the simplistic NPK (nitrogen, phosphorus and potassium) formula of conventional, petroleum-based fertilizers. Even if you add calcium and/or magnesium, like some "advanced" formulas do. Healthy soil has so many nutrients and micro-nutrients that *literally* we don't know what they all are and certainly not exactly how they all work together. Soil doesn't just hold up your plants, healthy soil is its own very interactive and very complex little ecosystem.

To reiterate a previous point: plants have hormones and a natural chemical make-up that we are still trying to fully understand (yep, just like us). But we don't fully understand it at all. Yet. What we *do* know is that these hormones and such help plants grow strong, produce better fruit and defend themselves (read: without chemical intervention) against both predatory pests and fungal diseases.

Yes, I totally get that you have used those bluish colored fertilizers for years and have grown all these wonderful, huge, prolific plants. Well, except the ones that just didn't thrive. But those aren't the ones you concentrate on, right? Or the ones that seemed to be a pest magnet. Or got a fungal problem. We won't worry about those...But those plants are the symptoms of your dead or dying soil. I know it *looks* nice and dark, but if your plants are not doing very well without that blue stuff and its' related laboratory-made brethren, then you can bet there is something else going on. Or, to be more accurate, nothing else going on.

What happens when you apply a little bit of the petroleum-based blue stuff is that your plants get a bit of a boost in some of the bigger nutrients and have a short period of increased growth and fruit or flower setting. Sort of like how you feel if you get one of those highly caffeinated "energy" drinks and then go and kick tail in your big presentation at work. Just like one energy drink (or two, for some of you) won't drastically affect your overall health, a little bit of the blue stuff won't drastically alter the nutrient make-up of your soil in the long run, but there are some strong indications that even a small amount is very detrimental to the microbes and other itsy bitsy critters that live in the soil and make up a vital part of this ecosystem. Most studies indicate though, that enough of them are left that the soil can recover in a relatively short time. If, of course, you don't use any more.

But if you don't apply more of this quick fix booster blue stuff on a regular basis, the effect on your plants is short lived. Just like your own productivity wanes as the effects of that energy drink wear off. And because the directions say so and because you want **more** and **bigger** flowers, plants and veggies, you apply more. Regularly. Maybe even just a wee bit more than the directions specifically tell you to because hey, if a little is good a lot must be great, right? (Take two, they're small.) And truthfully, the first season you do this, you will likely experience a certain level of success and be soooo very happy with your garden.

The second year, or maybe even part way through the first, you may notice that there are some problems. Some plants aren't doing well, maybe getting some fungus, maybe getting severely attacked by bugs. So you run down to your local garden center, describe the problem and come home with another bottle of chemicals (or two) and get to work spraying them. And still applying the blue stuff. Meanwhile the weeds are growing and the plants are giving you more problems and you are getting a headache from the bug spray. You apply more blue stuff, more bug spray and since you are now inundated with weeds, how about some carefully placed herbicide on those nasty suckers?

Please understand, what you are really doing is *killing* your soil. Yes, I know you were diligent in adding lots of compost and manures at the start of the season and your soil is visually appealing and feels perfect. But what it looks like and what it feels like are only a part of the equation. Frankly, a very small part. The microbes and bacteria that are naturally present in healthy soil (and compost) aren't things you can see with your eyeballs. Petroleum based fertilizers and the other man-made chemicals you apply to kill the weeds, kill the fungus, kill the bugs and fertilize the plants will also kill the microbes and good bacteria that your soil would otherwise provide the perfect home for. And *that,* my friends, is what is so very harmful about using those products.

When you use these products regularly, you are creating a downward spiral for the health and life of your soil. "Plant health" is discussed so much, but there is absolutely **no way** anyone can discuss plant health without also discussing *the health of the soil.* I cannot possibly emphasize this enough. Imagine the posture and voice and demeanor of the person who can influence you the most, multiply that times ten and then hear them saying: soil health *is* plant health. It is not possible to separate them. You already know this if you have a shed full of chemicals and a bit of frustration every time you have to buy more.

You also know this if you are an energy drink junkie. You know you can't sustain the level of energy the drinks give you. If you have ever tried to sustain that energy by having *more* of these drinks, you know it just doesn't work that way. More energy drinks just make you cranky and irritated and *less* able to function properly. You know without anyone telling you that you certainly can't live on energy drinks alone, you need actual *food and rest*. Well, your plants need food and rest too, and just because energy drinks give you a short term boost doesn't mean they are food. The blue stuff isn't really food for your plants and soil, either.

In the quest for bigger and better plants *right now*, we have forgotten what the true goal is: healthy plants that produce healthy food. Somehow, somewhere, "big plants" came to be equated with plant health and successful growing. Sorry, not true. Joel Salatin, modern day guru of the proper way to grow food, said something that totally pulled together all of what I know about growing anything and everything into one cohesive statement; he said, "An ant is the size of an ant for a reason." Isn't that wonderfully perfect? What that means is that some things are not meant to be as large as a house. Would you prefer your peas to be as big as a softball? Your green beans as big as a bat? Then why try so hard to make your plants so big?

What's so wrong with big plants, you ask? Big plants can be cool. When they are supposed to grow big. But plants bigger than they are supposed to be, particularly bigger *faster* than they are supposed to grow, *are* a problem. If you have ever been a fan of the world record-type books, you know what happens to the people listed as the tallest in the world for that year. They die young. Why? Because they grow too fast to grow properly. Their bones can't keep up. Eventually the sacrifices the body has to make to maintain that growth show up in forces so big, the body can't sustain life. The same thing happens to your plants when you try to make them grow bigger faster than they would naturally grow on their own: their infrastructure (their "bones") has to sacrifice something to give you those big plants. Given the proper time and nutrition and, of course, good weather, a tomato plant will keep growing and growing and producing and producing. Fifty foot tall tomato plants are not unheard of nor particularly surprising. But folks that wish they could grow a fifty foot tomato plant in one summer are what makes this idea a problem.

You won't get a fifty foot tomato plant with the blue stuff. There aren't enough nutrients in those laboratory-born formulas to sustain life over the long term. They are just an energy drink. Sure they can put vitamins in an energy drink, but does anyone think this means you don't need to eat food, too? Why? Because there are other nutrients, vitally important nutrients and micronutrients in food that simply cannot be made in a lab and inserted in a drink (or even in a pill, despite what too many years watching The Jetsons ™ would have me believe). So why do you think the plants that grow your *food* can be treated in a way you wouldn't treat yourself?

Adding petroleum-based fertilizers at the level recommended (And you know you will add even more. Plot spoiler: the manufacturers know that too.) greatly diminishes the microbial life in your soil. This microbial life is to your plants like eating food instead of just energy drinks is to your own body. The plants need these microbes to make hormones and such to attract good bugs and repel bad bugs. They need them to defend themselves against harmful fungi. Most importantly, they need them to impart all of the vitamins and minerals of the sunshine and the soil and the water and the air into their fruit so you can get those wonderful things into your body where they belong. Your body needs these things to function at its highest level. Period.

The reason you can use the blue stuff year in and year out and not totally kill your soil is because it generally has a season or two when it lies dormant and it can at least partially recover from the chemical damage. This is especially true if you, like so many gardeners, add some or a ton of compost and sheep poop and the like to your garden every year. In a normal garden, you shouldn't have to be adding so much every year; your garden should be getting nutrients from the natural mulches you use and homemade (not brought in a bag or truck) compost you generate from your very own waste. Healthy soil with healthy worms and microbial life will *generate its own nutrients.* Hopefully you just realized that the cost of this book just saved you potentially hundreds of dollars next spring. Pretty good return on investment, wouldn't you say? This is even before you read the research that places run-off from the chemicals used for suburban lawn and garden care as a *major* environmental hazard and even before you remember the most important thing of all for those of you who eat:

You know the difference between the way a tomato you get in the off season at the grocery store tastes and one you grow in your own garden? There is also a huge difference in taste between a tomato grown in a garden raised on chemicals and a garden raised naturally. A huge and absolutely lovely difference. And that, my friend, is worth not only the price of this book but we'll raise it with the "cost" of you turning someone else onto this extraordinary, virtually free process known as vermicomposting.

Worms and their associated microbe brethren are naturally present in healthy soil that has sufficient organic matter. In the normal course of soil life, plants grow and die and decompose. The worms and soil microbes feast on the decomposing matter. The bacteria and microbes in the worms' guts (and the guts of the other dead-stuff-eaters) turn the decomposing organic matter into tiny nuggets of nutrients and micronutrients that your plants, in their turn, need to do all the cool stuff I already mentioned. Worm poop is also a major source of humic acid, which is *essential* to soil health and not something you can just buy at the store. Kindly enough, their guts coat these little poop nuggets in a bit of a mucous-like material that allows all the good things in the poop to be released over time, so they are there when your plants need them. This helps the plant grow and set fruit all in the proper time. More importantly, they help the fruit set with the proper nutrients so that you, in your turn, can get them all into your body in the yummy package they were intended to be put in. It's all part of the process and **you** are the most important part in making sure that this process continues on schedule.

Harvest Time

Okay, okay, sorry if you already knew all that but so very many people don't or don't seem to so I had to put it in there. Now, off the soapbox and back to the poop:

The question always comes up in my classes, usually well before we get to that point in the lecture: *when* will I be able to harvest the poop? Everyone wants a number. Lots and lots of people will give you a number. They will say three months or they will say ninety days. (Yes, *I* realize these are the same thing; I'm not the one saying this, I'm only telling you what the nameless say.) Seems like that is what they always say. (They are usually also trying to sell you something.) I say: poppycock! I won't give you a number because if I do, that number will likely be a lie. At least at first. But I'll also tell you why I can't tell you an exact number and I think, if you have followed me this far, you will totally understand.

I have a friend, who shall remain unnamed but her first initial starts with Stephanie, who *mercilessly* made fun of me and my worms for about a year before she casually asked one day about maybe, someday, possibly getting a worm bin of her own. (This will also happen to you. A lot.) So I set her up with her own worm bin and explained the process. She did great for about a month, when she asked me how to get the poop to use on her plants. Seems she was under the impression that the bedding is where the worms lived and the food is what they ate instead of understanding that the worms would eat everything.

She brings me over to her bin and opens it up to show me how she is doing and shows me the little worm poop particles she could see all over her paper shreds. She points to them and asks if she just puts the paper with the poop onto the plants or if she has to pick it off and, if so, how? No, really, *pick it off.* Because I'm mean that way, I started to explain the "tweezer extraction process". She eventually caught on that I was full of poop myself and we got to the truth of the time it actually takes. Which is, like so many other things in life, dependent on many things. (And lest you haven't figured it out already: the whole bin will be turned into poop and tweezers will not be used at any time.)

Remember that when you are doing your bin maintenance, you are adding bedding regularly to keep the depth of your bin at an acceptable level. At some point *that only you can decide since you are the one doing it,* you will stop adding bedding and only add food. When this is going to happen exactly is a little up in the air, and also why "ninety days" is such b-o-l-o-g-n-a. The start of a new bin always has a lot of "it depends" attached to it: how many worms are in the bin, what level of maturity are they at, what is the level of decomposition of the food available to them, are the moisture and air always at the proper level, how big is the bin, how deep is the bin...All of these are crucial factors in calculating the "when" of harvesting.

There is at least one company in the U.S. that will, for a pretty high price I understand, let you "buy into" their business of raising worms. You pay "X" dollars and they send you the exact formula for bedding and food and worms and container size, you raise the worms and sell them back to the company. At least that is the story I get from people who have done it, and they are sworn to secrecy by the company so shall not be named. The point is that you *can* develop a formula to a certain extent. The problem is that you are dealing with living things. Living things tend to not always fit into a formula. Which is another reason to avoid pyramid schemes like this: they are run by living things that may not be entirely forthright.

So, you ask in total frustration now: when? I like to harvest my bins about a week or two weeks after I decide they are "almost" ready. Most generally in a bin-type of environment, this will happen somewhere between two and four months after you start. If it is your first bin and you are at the bottom of the learning curve, it could be as long as six months. Too aggravatingly non-specific? How about this: If I look into a bin and see that there is very little in the way of bedding left and mostly just pieces of eggshell and slower to decompose food bits like avocado shells and the like, I know it is time. Not time to harvest, but to stop feeding. I stop feeding a week or two before harvest to allow the worms to "clean up" the most of the remaining visible food bits. Because I also sell the resulting vermicompost, I generally stop feeding coffee grounds a few weeks before this. I don't like to sell coffee grounds to people, they can get those for free at their neighborhood coffee house. But if your vermicompost will only be for personal use, have no concerns about the coffee grounds, they are good for your plants, too. I also cut back just a little on the water for a week or two before I harvest. Not by too much, but slightly drier poop is way easier to harvest than wet poop.

About a week before you think it's time to harvest, make enough damp paper shreds to fill your bin all the way up again. Don't put them in the bin, just have them ready. This is also a good time to decide if you are going to need a bigger bin or a second bin for all of the extra worms you now have. If so, stop by the store or figure out how you want to make yourself a new bin.

I don't recommend that you stop feeding for longer than two weeks, though. Longer than that means that there may not be enough food left over for the worms to eat. Remember that you will have more worms than you started with because they will be mating and growing that whole time. If you stop feeding for longer than that, you risk losing some worms. Depending on how much bedding and how many food bits are in the bin when you feel it is almost ready and depending on if you pick one week or two (or three because your friend had free tickets to a wicked cool concert and you went to that instead of harvesting your worms like you were planning on doing or some such), you could lose a few worms to starvation doing your harvesting my way if you don't stay on schedule. So make sure you either stay on schedule or give your worms some food if you are going to overshoot the two week mark.

My way to harvest depends on you spending at least five minutes a week in your worm bin so you can be regularly familiar with the conditions and develop a feel for what is going on. Sorry, I don't do formulas. I don't think they are reasonable or entirely possible. You will also see, once you read some of the other ways to harvest, how this is the easiest and most efficient way, even if it means you do have to pay a little attention to the timing. The timing which you decide all on your own, not by a calendar or by a formula. All of the OCD people just threw this book across the room. But seriously, you may not be able to envision what I'm talking about right now, sitting on your couch reading this book. When you get to that point, though, it is much easier to see.

For the rest of you I will say this: if you are that unsure, keep feeding a little bit until the very day you are actually ready to harvest, but make sure your food is all in a big lump, where you can reach in and pull it out and put it in the new bin when you get a free day after that concert with your friend. Then it becomes food in the new bin and you can harvest on even less of a schedule.

Don't over think the harvesting too much. It will make more sense to you when you are actually doing it than it may when you are just reading about it. If you are raising worms to sell their poop, you will want a finished product that looks nice. If you are just using the compost for yourself and maybe a neighbor, then you don't need to worry about there being lots of bits and pieces of eggshell or a little bit of paper in the compost. It will work exactly as it is supposed to even with these extra bits and the extra bits will eventually decompose and get eaten in your garden just like they would if you leave them in the bin.

One thing I cannot emphasize enough, but, when you read the next sentence you will realize just how easy it will be to remember...Worms don't like to live in their own poop any more than you would want to. Yeah, I agree, ewwwwww—but something you will now never forget. Yes, you can get them to live in their own poop for quite awhile. There is, however, a limit even for them. So if you did actually go to that concert and skip the harvesting you were supposed to do that week, don't wait for too many more days to roll by before you get to it. Your worms will start dying off if you do. I can only pray that is enough incentive for you. You cannot postpone harvesting forever, at some point you *will* have to harvest the worms from the poop and start a new bin.

As you Sow...

How to Harvest

Now that the "when" part is taken care of (or as much as it possibly can be) let's talk about the "how". Just like there are other ways people have found to raise worms, there are also many, many different ways to harvest their poop. Just like I told you about raising worms, there are lazy ways (my take: more efficient) and there are ways that are generally successful, but that either take more time or take more effort. Pfffft! We don't have all that time and have other things to give our efforts to, like using the poop! So first I'll tell you what I have found to work the best, and then I'll cover some of the other methods and why I don't think they work as well.

Harvesting one bin the way I teach will generally take about an hour to an hour and a half total, depending on how efficiently you work. First, find a spot where you can lay out either a plastic tarp or two large garbage bags that you cut open and lay out flat to make a large square. You can certainly do this on a bed sheet or some such, but something made of plastic works better and is easier to clean. You can also use an old oilcloth or plastic tablecloth or old shower curtain if you have one handy. All will work equally well, though the thicker plastic will be easier in the long run. After you get your tarp (or whatever), lay it out somewhere you can work where the dog or cat (or the neighborhood birds) won't get into it while you are working on it. Now find a lamp that you can use that will put extra light over the area of the tarp and an empty bucket or other container you can use to put the finished vermicompost in.

Now, go get your bin. I like to wear gloves during this process so I don't have to wash my hands so much in such a short time. As long as your compost is not too wet (and you'd never do that, would you?), cloth garden gloves will work just fine for this. When you are ready to begin, dump all the contents of your bin in the center of your tarp. Make sure you dig out the corners of the bin and get all the worms and everything wiped out gently with your hands. No need to clean it if you are going to start another bin unless you want to because you are just that way.

Set the empty bin to the side for a moment. Take the big old pile of worm poop you just dumped and just take a minute to admire it. Wow! That used to be a whole bunch of *garbage*! You did good.

Take the big pile and make a bunch of smaller piles or low rows out of the poop. Doesn't matter how many or how long or wide the rows are. Just make sure there are at least six inches of space in between the piles or rows. More than six is fine, too. Turn the lamp on and make sure the lamp light covers as many of the poop piles as possible. (The lamp does not need to be super bright, just regular bright. You aren't trying to interrogate the worms—Where were you on the night of July 12th?!?!—you are just trying to take advantage of their photophobia.)

Now go away for about ten minutes. This is the perfect time to pick up the empty bin and go fill it up with the damp paper shreds you made last week and new food to get it ready. This is also the time to decide if you are going to need a second bin, if you weren't sure before, because you just realized when you dumped out your bin that you have too many worms for just one. If so, get that started now, too.

Ding! Ten minutes (give or take) have passed. Now go put your gloves back on (if you decided to use them) and go to your poo piles. Start with one and gently brush back the poop until you see a worm or worms. They will be easy to see because they are pink and the compost is very dark brown verging on black. Once you see a worm or two, stop. Scoop up the worm poop that you brushed off and put it in the designated bucket. Gently make the pile you just worked on a little smaller by reforming it into a "tighter" pile. By tighter, I don't mean actually tight, just closer together. Move on to each successive pile or row until you are done. If you have a lot of worm poop to harvest, you can probably just go back to the first one and start the process all over again. If not, go away and read a book for ten minutes, then come back and remove more poop until you see a worm, and so on and so on. You can combine piles if they get too small if you like, it doesn't seem to matter much time-wise if you do or if you don't.

Because your worms are photophobic, they will burrow down more deeply into the piles as you go. Eventually, you will have very little poop left to brush away and instead of a pile of poop, you will have a small pile of squiggly worms. You won't get all the poop off the worms, which is totally fine. When you have removed all the poop that you can, scoop up the worms and put them into the new bin or bins you made ready at the beginning of this process.

There will likely be both babies and some potworms hanging out on the tarp after you scoop up the worms. Very, very carefully, with either a bare hand or a latex-gloved hand (for this part the cloth gloves don't work quite as well), pick up the baby worms with a light sideways "scoop" of the finger. Don't try to grab them by pinching them, you will more likely squish 'em. They are very delicate little things. Remember, the baby worms will be pink and the potworms are white. The potworms, well, you can choose to save them or not. That is entirely up to you. I usually just take the tarp out and lay it dirty side down in the shade and they find their way off the tarp and into the ground just fine.

I mentioned before that baby worms are not so bright. They also can't move very fast. While all the other, larger

196

worms nosedived away from the light, so did the babies. But because they are just babies, they can't move as fast. Because they are so small, you also may not have seen them when you scooped up the worm poop from the piles. This means that there will likely be a few itsy bitsy baby worms in the compost you removed from the piles. Because of this, your new worm bin will have a slight delay in the expected population increase because some of the babies are in the vermicompost you just harvested. The same holds true for the cocoons. A very large number of the cocoons will be in the compost.

Unfortunately, all harvesting methods will result in *some* level of baby and egg loss. Well, not "loss" really, since they will be in the compost and you are going to use the compost. I have tried just about every way there is and for home bins, I have found that the "dump and sort" method results in a minimum of loss. "Acceptable loss" if you will.

Some people will have a really, really hard time with this. In part, because they will want every single one of the worms in the bin where they can eat the garbage and grow up to be big worms that make more worms. If this bothers you that much, you can take the compost that you harvested from the piles and go through it a second time, under a much brighter light, searching through all of it by hand to pull out the eggs and the babies. Prepare for your neck and back to hurt. Obsessive worm-woman I am, I must admit that I've actually done this before; though not often. It is not pleasant and the volume you get for the effort will always be disappointing. It certainly can be done, I just don't recommend it.

Keep in mind, too, that if you are going to use the harvested compost right away in your garden or lawn, a lot of those babies will live on just fine. Not to mention the worm eggs will hatch and those worms will likely be okay as long as there is something for them to eat and sufficient moisture. Usually not a problem in most organic gardens. These "leftovers" will actually help to increase the worm population in your garden *naturally* so they can live there and provide poop to your plants the whole season long. It's like bonus poop!

If you are going to store the poop instead of use it right away, like because your compost harvest occurred in the middle of December, a lot of the babies will likely die off over time, but the eggs will remain viable as long as there is sufficient moisture so they don't dry out and it is not so cold they all freeze. Most of the babies will end up at the bottom of whatever container you are storing the poop in because the bottom will have more moisture to keep them alive. I keep my stored poop in a cool place, slightly covered but not totally covered. I usually just place the lid on cockeyed and give it a little bit of water every now and again to keep the bacteria and such moist. You will want to keep it about half as moist as the moisture level of your living bin. If the stored poop has no air it will start to turn anaerobic and get a little stinky. This occurs especially fast if it has little or no air and is stored somewhere really hot, like your garage in July. So make sure you leave the lid off or at least half off and don't store it anywhere that it will get too hot. If it does get stinky, just give it more air and it will correct itself shortly; it doesn't mean the poop is ruined and unusable.

As you can imagine that whole harvesting process took just a little bit of time and now you have your new bin(s) started and the whole rest of the day to work in your garden or read a book or write a letter to me saying how much you have enjoyed the whole process! Or you can keep reading about some of the other methods of harvesting so you can be really happy that you didn't use one of those...

Other Methods

The most common method of harvesting described, besides the one I just told you about (which isn't something I invented, mind you, it's just something I do), is to simply add new bedding and food to half the bin and let the worms migrate themselves to the new stuff. All you do is move all the finished compost to one side, add new bedding and food to the other side and wait for the worms to migrate from the old stuff to the new stuff, then scoop out and use the old stuff and fill that part back up with new bedding and fingersnap! You're done.

Sounds like the ideal lazy-worm-girl way, doesn't it? But remember, I'm not just lazy, I'm impatient. While this method does work to a certain extent, it doesn't work *quiiiite* the way it's advertised to work. When you read about this process, people generally say the worms will move over to the new bedding in about two weeks. You may be surprised (but hopefully not by this point) to find that almost the exact same wording is used on every website or in every book you read about this technique. Almost like it, too, was copy/pasted a zillion times over. The problem with this method is the "finished" half. Although worms do like to have lots of different food around to choose from, they also like to hang with their pals and be in an environment that is familiar to them. They also only like food that is decomposed. Like the food bits in the old half. It takes them quite awhile to decide that the new food you put in the new half is worth the trip. The "two weeks" this is supposed to take turns out to actually be more like three to four weeks. During which time you will still have to do some regular maintenance in the new half and make sure it is still in good condition *and* still keep it separated from the old stuff.

Yes they will migrate to the new bedding, but just like when you add new bedding to your bin during your regular maintenance, your worms will go to both the new bedding and the older, more finished compost. They won't ever completely leave the old stuff no matter how yummy the food is that you use in the new stuff. To add to the problem, the baby worms will be birthing themselves all the time in the old stuff. With this method the problem of baby worms and cocoons remains the same. They will be stuck in the old stuff. The babies simply don't have the smarts or the strength to move to the new stuff. And the eggs? Well, they can't move on their own at all.

If you want to wait until the babies get big enough to move over to the new stuff, good luck. They will eventually do so, but at the same time all of the cocoons will be hatching new babies and you end up waiting for the babies to all grow up and move over for just about forever. Which means if you want to use this method and also keep all of your worms, you will still have to pick through the finished compost to retrieve the worms that like the old home better. My philosophy is that if you are going to be doing this anyway, you might as well do it all at once.

Another method that is very similar involves putting a screen in between the new and the old halves. I must say that I have never figured out the reason for this except maybe to help keep the two sides separate. If you want to try this, don't use window screen as some instructions suggest, it is too small for most worms to squeeze through, especially the night crawler size. Use the type of heavy screen more commonly known as hardware cloth. It is sold in rolls at the big box hardware stores and by the roll or by the foot at the smaller stores. I would guess that the eighth inch hardware cloth would be better than the half inch or the quarter inch, but again, I don't know why you would bother at all.

Another take on this same method does involve a screen and in this case rightfully so. It is the exact same concept, except you put a screen on top of the older compost and put the new food and bedding on top of the screen. The theory is also the same: the worms will be so enamored with the new food that they will scootch their way through the screen and into the new bedding to get to the new food, then all you have to do is wait a couple of weeks, lift out the screen with the new bedding and food and, theoretically, all of your worms, scoop out the old compost and put everything from the screen back in the bin to start all over again. The exact same issues happen with this vertical migration as with the horizontal migration.

Commercial or homemade stacked bin systems like I described waaaay back at the beginning also work by the vertical migration method. Exactly the same problems occur in this type of system as in any other migration system. Babies and eggs get lost in the process the same way and for the same reason. Part of why people like these stackable systems is that, because of the number of bins in the stack, by the time you do get to the top bin, most of the worms in the very bottom bin will have migrated up. Because you just stack a fresh bin on top every time the bottom one is finished, you don't have that foot-tapping wait for the "two weeks" (as if) it takes for the worms to migrate so you can harvest the poop like in the previous two methods. So it feels like it works just great. There will still be babies and cocoons though, not to mention several heavy bins you have to move to get to the ones on the very bottom tray so again, I don't know that this is the best investment.

"Flow through" bins are another option to think about, but maybe not until you have a few bins under your belt and want to try a "better" way. Some folks swear by a flow through system. There are lots of plans for these on the internet. The cloth "sack" systems you find sometimes are the simplest version of a flow through. Lots of commercial worm farms use a flow through system, too, though on a much larger scale and more permanent set up than the sack type. Essentially any type of flow through system is set up from the very beginning of the bin life. The system is set up in a way that allows you to harvest smaller amounts of the vermicompost at regular intervals from beneath the bin, either by opening the bottom up a bit (in a sack system) or scraping the bottom via a bar set near but not at the bottom that loosens up the bottom bits of compost and lets them flow through and out the bottom (hence the name). Guess what? Same problems with these types of systems as far as the babies and cocoons go, but generally fewer. The reason I don't like these systems for personal sized bins is because they don't give you a lot at one time for you to be able to use when you want. I like to have all my poop where I can get to it when I want it. Um, I mean my worm poop. Don't want any confusion on *that* point for sure.

Sifting

Another way to harvest is by sifting. Sifting will give you a much nicer looking finished product. I use the hand sorting method at certain times of the year when the Colorado weather doesn't cooperate with my harvesting schedule for my indoor bins. I never hand sort the outdoor worms because the volume of those areas is just too large. Sifting is the harvesting method of preference for most larger scale operations, but you can do it even with one or two bins, too. The difference lies in the type of sifter. A commercial facility will have an electric sifter. If you decide to use a sifter, you'll make it yourself and most likely power it yourself.

You can sift before you get the worms out, or you can sift afterwards to have a nicer end product. Or you can not sift at all. It isn't required for home bins, it is just another option. Sifting, like everything else, has its benefits and its downsides. Choosing to sift will depend a lot more on how you feel about the way your end product looks and feels and on your energy level than on anything else.

There are an infinite number of ways to make a sifter. Essentially it involves some type of hardware cloth or heavy screening with some type of frame. (Again, window screening of any type is not going to work because the openings are too small.) A sifter works because most of the worms will stay in the sifter, along with the larger pieces of food or otherwise unfinished stuff from your bin (this larger stuff is called "overs" in the compost business). Depending on how you make your sifter, the worms will stay in the sifter. More or less. The rest will fall through with the compost. And therein lies the rub.

Sifters are great to sort through everything and to separate the worms from their poop if you have a lot of bins, or even one outdoor windrow. If you don't have a lot of bins, then generally they are just for a more aesthetically appealing end product and are not really much of a time saver when harvesting the worm poop. Some worms, especially the smaller ones, will still fall through no matter what type of screening method you use. Cocoons, too, of course. Some of you will prefer a sifter no matter what I say so I might as well give you the tips and tricks now and save you the trouble of making something that just isn't going to work right.

The easiest way to make a sifter is to make a pan out of the hardware cloth. I've made the pan out of straight hardware cloth alone folded up about five inches or so on each side and sewn into place with a bit of wire; it basically looks like a shallow box made out of screen. Most people like to make it with the hardware cloth stapled to the bottom of a wooden box frame. A frame makes the sifter much sturdier and more long lasting, but also heavier. If you do this make sure it is also designed so it is comfortable for you to hold.

A single sifting with the ¼ inch cloth is fine and will give you a good looking end product, but the 1/8 inch will give you a very fine product and will also catch more of the worms than the ¼ inch will. Which means, you guessed it: if you sift you will still have worms in the compost that need to be picked out if you want them all to go into the new bin no matter what sized screen you use. Are you sensing the theme here: no method of harvesting is foolproof. If you want to prove me wrong on that, by all means *please* (pleasepleasepleasepleasewithsugarontop) do so. But only if you send me the plans for the system. (There's always a catch.)

Sifting with a homemade sifter will involve making sure the compost is as dry as it can be without doing any harm to the worms. Sifting wet compost is just not possible and is merely a recipe for extreme frustration. I recommend dumping and sorting the worms on a tarp before sifting; I know it doesn't seem like it, but it will make things go faster. All you do is take a couple three of handfuls of compost, put it in your sifter and shake it over a tarp. Put the stuff that doesn't go through the screen back into your new bin and the stuff that falls through will be all nice and purty and ready for you to use as you see fit.

If you didn't take my advice and do the whole dump/sort shebang with the compost beforehand, you will find that your big worms mostly stay in the sifter frame and the smaller ones went right through. Especially if you used quarter inch screen. So now you will have to sort through the compost anyway. See how it makes more sense to just do it all at once? You will also find, especially if you have more than one or two bins, that dang, your arms get *tired*.

I had the same problem. Due to the volume of worms that I had at the time, all that shaking was getting to be a pain. This was when I was first starting out and didn't have any outside worms, but a whole heck of a lot of worms in bins. I needed a solution that did not involve all that shaking. Commercially made electric sifters work great, but usually start at about a thousand dollars even for a small one. Yikes, eh? That's what I said, too. What I eventually did is convert a five gallon bucket into a very nifty sifter. The total cost for this project was less than twenty bucks and I'll describe it here in case you do get worm fever and end up needing a more efficient sifter yourself.

You will need somewhere to anchor or "hang" the sifter. I use two sawhorses, but if you don't have that, you can do it over a deep wheelbarrow or some such. Use your imagination, but make sure you have something relatively sturdy. Even two step ladders set up on each end will work. Ideally you will want this set up so the sifter will rest at or just above hip level. Find or buy about a three or four foot section of two inch pipe. I used PVC pipe and it has worked well for a very long time. Still does, actually. You will need a way for the pipe to stay still on your anchors. I have C-shaped bars screwed into the top of my sawhorses. A rope or bungee cords through the pipe will work, too.

Caulk or glue the lid onto a round, five gallon bucket; you want it to stay tight. Drill out two inch wide holes into the center of the lid and the center of the bottom of the bucket. With a reciprocating saw, cut "windows" into the long sides of the bucket, about four inches from each end and about six inches wide all the way around the bucket. Leave at least two inch strips of the bucket sides in between each "window'. Drill small holes near the top and the bottom of every strip and in the middle. This all works best if you draw where you want to cut on the bucket first with a permanent marker. Drilling the holes in the strips before you cut the windows out will also give you more stability. This is all much easier to understand when the materials are right in front of you, promise.

Decide if you want to use quarter inch or half inch hardware cloth to sift. Cut the cloth to fit all the way around the diameter of the bucket with about a foot overlap. Strong scissors will cut the cloth with a little muscle behind them, or use a pair of tin snips. I like to make the cloth a couple inches longer on all sides and fold it under to keep the sharp edges away from me. They hurt! Use wire or zip ties to secure the screen to the holes in the strips, leaving the overlap unsecured. The overlap is your opening into your sifter and you can easily secure it with a couple or three hinge pins (about ten cents at the local hardware store).

Now all you have to do is put the pipe through the holes in the lid and the bottom, secure it to your anchors, set a tarp underneath, put the compost into the sifter, secure the opening closed and turn, turn, turn. The cool thing about a round sifter is that very few of the worms will go through into the sifted compost. Worms are long and thin and the centrifugal force will keep most of them in the sifter while it is moving. (No, not G-force type of turning, just regular, light turning works just fine and will keep your worm poop from flying all over the place.) You can scoop the worms out with the rest of the overs when you empty the sifter. It's really easy, but not necessarily worth the effort if you are only going to have a bin or two of worms.

There are, as I mentioned, also commercially made electric sifters available to purchase. If you are super handy (or know someone who is and who also likes you a whole heck of a lot), you can make one for substantially less money than it costs to purchase one already made. These are wonderful if you have a ton of worms, but for the vast majority of you, the dump and sort method will be just fine and you will likely find that you enjoy doing it!

Poop 101:

Basic to Advanced Poop Usage

Woohoo! Finally, eh? So many people want to learn how to raise worms solely to help keep this gorgeous planet clean and to do something to help run the landfills out of business. But so many more do it for the poop. The whole poop. And nothing but the poop. Well, either way, you made it. Congratulations!

I've already covered the wonders of worm poop and what it will do for everything green. You will just have to make some and use it and see it for yourself. And *taste* the difference in your produce. Oh!

There are really no wrong ways to use the poop. It won't hurt the plants you are growing in any way. It won't burn the roots or the leaves. Ever. You can't use so much that your plants will die. You can't hurt your soil any by using it, either. Quite to the contrary! You can use it when you plant your seeds, when you plant your seedlings, to amend the mix when you plant your trees, as a top dressing, as a side dressing and as a compost tea (liquid fertilizer). You can also use it as a foliage spray to increase plant growth and combat fungus and predatory insects and disease. And because the poop can hold two to three times its weight in water, you won't have to water as often. All without ever hurting your plants and trees, or the soil, or the ground water, or you and your family, or the planet. And it used to be garbage. Go figure.

Heck, I'm pretty sure you could even eat some and suffer no ill effects, though it might not be so yummy in your mouth. Would you want to try that with the blue stuff or any of the other chemicals you have in your garden shed? How about we use *that* for a test of what should be put onto plants: if you couldn't conceivably eat it with no ill effects, you shouldn't put it on your plants. I like it!

Let's start by talking about how to use the worm poop in its solid form and then we will get into the wonderful world of compost teas...

I've read a lot about raising worms and using their poop. I've mentioned that I find many articles and blogs and boards use almost or exactly the same words to describe various aspects of worm farming. Does this mean we've got a passel of cheats and plagiarizers on our hands? Not at all. What it means is that there is only a small body of knowledge available in regard to raising worms for composting purposes. Considering how important this endeavor is, one would think there would be more than just a small handful of books and pamphlets around. But there just aren't.

Some soil scientists may study worms, but they are studying them in their natural environment and in the way they interact with other creatures in the soil, they are not studying how to raise them in a confined space. Some commercial vermicomposters and even some municipalities in the world raise literally *billions* of worms, but that doesn't mean they are doing a lot of experimentation or that they have time to spread the worm, um, the *word* about how they are doing it and the problems or improvements they have encountered. There are even people around who have had a worm bin or several for decades. Decades, people. But for some reason they are keeping their tremendous body of knowledge to themselves. For shame.

This is probably why you read the same sentences over and over again on the internet and in the few books there are about raising worms. Also, of course, because so much of it works. Hopefully by this point you have realized that I'm not just trying to share what works, but what works *better*.

One of the things you will or already have read about over and over in other places is how to use the poop. Pretty much everything I've ever read uses the exact same "formula" about using it: mix the poop in with your soil at a rate of twenty to forty percent of the mixture. I've read this a zillion times and even tried it my ownself. Does it work? You betchca. Did I find a better way? You win again.

When I first started out raising worms, I started out with just a couple of pounds and I had my share of problems. I received my first worms in February and by May, which is garden planting time here in Colorado, I only had about ten or so pounds of poop. Now that might seem like a lot, but bear in mind I live on 35 acres and my garden area was about the same square footage as my house. Ten pounds just wasn't going to make up twenty to forty percent of anything.

To make matters worse, the acreage I bought hadn't exactly been nurtured along by anyone who could have even remotely been described as a tree-hugger. "Sustainability" was not in the vocabulary from what I could tell. I also chose to put my vegetable garden in the most convenient place for me. Weeds literally had a hard time pushing through the concrete-like dirt. To illustrate: my friend tried to till it for me to loosen things up. The tines on the tiller didn't actually bend, but they did actually leave *skid marks* on the sand embedded clay that was my "garden". I'm stubborn. That is where I wanted the garden so that is where it was going to be.

I loosened the areas I could with water and a shovel. I had just enough money left over from all the house repairs to buy a few bags of mushroom compost to add some organic matter plus my ten pounds of worm poop. I put one very small handful of worm poop and a couple of handfuls of mushroom compost in with each seedling. Tomatoes, peppers, cukes, herbs, the obligatory zucchini plant. Nothing extraordinary. By September? Huge tomatoes, over six feet tall (not pruned), cucumbers that climbed over the fence and offered their sweet fruit to my dogs, more peppers than I could eat and enough basil that I actually had enough left over to sell some of it.

But what if it was just the mushroom compost that caused all the phenomenal growth in soil that was previously practically cement? Nope. And that isn't just faith. I planted some morning glories to climb up some old chain link I'd hung on the back of my loafing shed. Three rows of fencing, three packages of seeds. Being practical as well as stubborn, I used the worm poop on the plants that feed my belly before the plants that feed my eyes. Ten pounds doesn't go very far even when you are using only a little bit for each plant. I ran out. I ran out just before the third packet of morning glories was opened. I did, however, have more than enough mushroom compost.

So the last packet of morning glories got no worm poop, only mushroom compost. I am in no way about to disparage mushroom compost. I love it. But that last packet of seeds? They grew. But they never got much higher than four feet tall and their leaves never got much greener than a ripe lemon. My worms made enough poop that I was able to give everything some tea later in the season, which I did use on the other plants and which I'll tell you about in a minute, but I mostly left that section of morning glories alone, as a testament to the wonders of the poop. The yellow leaves and small, sparse blooms were a huge billboard in favor of the poop over the plain compost compared to how the rest of the garden looked. Yes, I do have the photos to prove it.

Why do I tell you this story? For a couple of reasons. For one, and bear in mind that I do sell worm poop so this is totally *not* in my best interest: I think that the "twenty to forty percent" that you have been reading about might be a slight exaggeration in most situations.

I say most situations instead of all situations because there are some exceptions to prove this rule. First of all, I made sure I used plenty of organic matter (leaves and grass and such)around the plants after they were planted. As the mulch decomposes, it feeds the worms that grow from the eggs that were in the compost when I put it in there. (Remember, you don't put worms in the garden, right? Their populations need to grow naturally based on the food that you have available for them or they will just die or leave.)

I'm able to only use a little bit on my vegetable plants because I make sure that there is plenty of worm food available throughout the season. Even though I only used a little, the combination of cocoons in the poop and the decomposing organic mulch serves to feed the growing worms and soil microbes that will in turn make more of their own poop and lay eggs and keep on feeding the plants all through the season. Understand? The small amount of initial worm poop gives the plants what they need to get a good start and, as long as I keep the soil conditions right with moisture and decomposing organic matter, the whole cycle will keep going on its own, almost like a gigantic outdoor worm bin. All I am is the steward and I save a ton of money by not having to buy anything else to add to the soil or to combat any problems with bugs or fungus. (I did have a wee problem with squash bugs on a pumpkin plant that I bought from a local store. I'm positive the bugs came with the plant and after that plant was eaten to bits because I found them too late, the bugs went away. They never went to the other plants, even though there was a spaghetti squash plant and some cucumbers less than five feet away.)

There are times, though, when you don't have the advantage of being able to mulch. House plants, container gardens, seedling trays and the like are not usually mulched. When I mix up soil for planting in a container or for my house plants, I add about fifteen to thirty percent worm poop to the mixture. I add more than I do in the vegetable garden because, in the absence of a steady food source, there is no way for there to be more poop made as time goes by unless you are able to provide food. Difficult to do in most container plants and I don't know too many people who want to mulch their indoor plants with old leaves. The addition of the worm poop will also greatly increase the ability of your container plants to retain moisture, which will cut down on the frequency with which you will have to water. Yes, there will be cocoons in the worm poop and yes, some of the worms will be born and yes, because there is no food for them they will eventually die off. Don't panic, it is a small sacrifice and you will be rewarded with stunning plants.

Remember that the mucous from the worms guts coats each piece of poo in a neat little "shell" that breaks down over time. Nature's way of making time release fertilizer. In six months or so, you will either want to add more poop as top dressing, to keep the container plants well fed, or you can choose to start watering with worm tea instead of plain water. We'll get to tea after we are done with using the poop in solid form. In any case, plants in a container will grow like mad with the addition of worm poop and you will be so crazily happy with the results you may end up living in a jungle of house plants.

You'll notice I mentioned seed trays at the start of this part but didn't mention them when I talked about adding worm poop to your container plants. As you may have guessed, I did that on purpose. I mentioned before that the "problem" with worm poop is that it does not destroy all the seeds like proper hot composting does (which is why you don't add weeds with seed heads to your bin, right?). People like me like to grow most of their vegetables from seed. If you have never done it, give it a try as it is one of the most wonderful miracles we get to regularly witness and participate in. Plus, you get to choose the varieties of fruits and vegetables you eat instead of letting some complete stranger choose for you.

I've found that most seeds from your frozen food waste don't sprout in the worm poop. Squash and pumpkin seeds, tomato seeds and grass seed seem to be the main exceptions. Grass, squash and pumpkin seedlings are pretty obvious when they come up and can be plucked out of the seed tray very easily. Tomato seedlings, though, can look a lot like other seedlings at first. As such, I generally don't use any worm poop in the seed starting mix when I'm growing tomatoes and peppers and eggplant and tomatillos and the like. Even some flowers like zinnias look too much like baby tomato plants until they get their true leaves. This is a problem because you don't want something you didn't plant coming up in the seed trays labeled as what you did want to plant.

So what do you do if you want to use the poop for your seedlings? Use tea. Your seedlings will benefit greatly, possibly even more than your mature plants, from the boost of the complex plant nutrients and microorganisms available in the poo. It will help them start out strong and healthy and they will better be able to grow and produce great food all through their lives as a result. I do recommend adding about ten to fifteen percent worm poop to whatever seed starting mix you use for the rest of your seedlings; but for your tomato seedlings and the like, I recommend only using either tea to water them or adding the poop as a top dressing after the original seeds have sprouted. Ten to fifteen percent is far less than what is usually recommended. I have found though, that this is more than enough to give them what they need to get growing well.

Some people think that if they sterilize the worm poop by baking it at a low temperature, this will eliminate the seed problem. They are absolutely right. Sterilizing the worm poop will take care of the problem of some of the seeds in the poop sprouting. It will also completely destroy all the yummy bacteria and microbes that make the poop work so well in the first place so don't you dare do this. Freezing your worm food will take care of the vast majority of the seeds, so please keep your poop out of the oven.

The ways to use the poop I just described might sound like too much "work" or too much to remember at first glance. And it certainly doesn't sound like the advice of a self-professed lazy gardener. Au contraire, mon ami! Doing this properly when your plants are first growing will save you a ton of time later simply by giving you healthier plants from the git go. But remember also how much time and energy you are spending mixing in bags and bags of whatever you like to buy every year to add to your soil and then buying and applying the blue stuff and then the herbicides and fungicides and insecticides and what not? You learned somewhere along the line that this was "the way" and you can unlearn it and relearn a better, easier, cheaper way in just one season. This really is the lazy way in the long run and it will only take you a season to notice the difference and to get the ball rolling so that your garden can become the self-sustaining little ecosystem it was meant to be in the first place.

After your initial planting, regardless of what or where, you will probably want to add some more poop, especially the first year you get started but even in subsequent years to give your plants a natural booster. Or you may already have plants or trees growing that didn't get planted with worm poop in the first place and now that you have all this cool new knowledge, you want those plants to get all the benefits, too. No worries. Worm poop is about the easiest thing to use, too. In solid form, you don't even have to worry about mixing it up with the surrounding soil. All you have to do is put some on the soil around the plants, either all around or in one place, really it doesn't matter much. A handful or two per plant is all you need. I even only use a few cups on my trees as top dressing (under the mulch). For outdoor plants it is definitely much more beneficial if you can also use some kind of natural mulch around your plants, not only to help keep moisture levels steady, but to provide food for the soil and worms as the mulch decomposes. I like to use poop tea on my plants and trees sometimes, too, and we are so very close to learning how to make tea so hang on.

As I mentioned before, adding worm poop to your garden plants will also help increase the worm population in your garden *naturally*. Particularly if you are diligent about adding natural mulches so they have food to eat. More worms in your garden will also give you the benefits of more active, beneficial microbes. They also help mix and aerate the soil, so your plants roots can have oxygen without drying out. Worm poop will reduce your water needs due to the increased ability of the vermicompost to retain moisture. All of this is super cool and you would never want to do anything to mess this up, right? Right.

Which means that, in addition to no longer using artificial chemicals in your garden, you must also commit to no longer running that bad boy tiller through the garden every year. No, really, I mean it. Without going into all the details about how your soil is already *in* the order it is meant to be in and not meant to be all mixed up top to bottom, let's suffice to say that running that tiller through a worm-filled garden bed is just the same as running the worms through your blender. Actually, it's worse.

I know you use the tiller every year to "loosen up the soil" so your plants roots can go where they need to go. The problem is that this is a myth. Sure, the tiller will loosen up the soil as deeply as you can get it to go. But then what? What happens to the area under the deepest point the tiller will go? Well, I'll tell ya. It gets compacted. Yep, you read right. The very machine you are using to loosen up your soil is also responsible for compacting your soil. True story. Imagine your garden soil in cross section. Now imagine the tiller going along the top. While those top few inches are whirling around like food in a blender, everything underneath the tines is getting more and more smooshed down. Doing this every year, or worse, twice a year, is actually doing more harm than good for your plant roots. Let someone who hasn't read this book yet buy your tiller and use that money and all that extra time to take your best pal out to dinner instead. Tilling should only be done once, to start an area out for planting and its not always necessary even then.

Everything I've mentioned about using the worm poop so far, and until the end of this book, involves using it as only a percentage of the whole soil mixture. When you look at worm poop, you will notice that it looks and smells exactly like the most gorgeous, dark, yummy soil you have ever seen in your life. So why then, would you not want to just use straight worm poop instead of poop mixed with dirt?

Well, you can. If you want. But you won't be as happy with the results. Remember when I said before that we know more about how the human brain works than we know about what is in soil and how it interacts with the plants that grow in it? I'm not just talking to be talking, there is a point to every story.

Lots of little studies have been done to try to determine the exact ratio of worm poop to soil that is most beneficial to your plants. This little experiment has happened a zillion times as science projects at colleges and high schools and junior high schools all over the world. This is where that original "twenty to forty percent" number that you read over and over again apparently comes from. All these little studies take "X" kind of soil (a bagged mix, a made mix, dirt straight from the ground, etc.) and mix it with worm poop at whatever ratio the person conducting the study determines is best. They take this mixture and some of the same soil without worm poop and then some straight worm poop without soil and grow the same kind of plant in each. Then after such and such a time has passed, they record the plant growth and root growth and sometimes even total water usage.

The mixture of poop and soil wins hands down over either the soil alone or the poop alone. Every single time. So yes, you *can* grow plants in straight poop or, as you already know, in straight soil. The thing to remember is that a mixture of the two works like gangbusters compared to either one alone. The exact "why" of this, though, is not totally known. What is known, what there is zero doubt about, is that they work better together than either works alone. Maybe someday they will figure out exactly how soil works and how everything in the soil works together to help plants grow. In the mean time, though, you can still use the knowledge we do have to benefit your plants and the environment and ultimately to benefit the earth in general. And you can use my personal experience growing many more plants than any science fair project to know that yes, in most circumstances, you can get away with using much less than forty percent.

There may come a time when you find that you have more worm poop than you do places to use it. This is most likely if you live in an apartment or town home with limited green areas available for you to use. In this case, you have several options. First of all, you can totally sell your excess on Craig's List or a similar site. But better than that is to give it as a gift to someone who isn't familiar with how beneficial it can be. Yes, I'm sure by the time you have "too much" poop, all of your friends will have already heard all about it. But there is nothing like seeing to lead straight to believing, so give them enough to get a couple of their plants through the growing season. Tell them to just use it on some and do what they normally do to the others. They'll see the difference for their ownself and pretty soon they'll be knocking on your door, asking to borrow this book. My advice when you reach this point: make them get their own copy! (wink wink) If all of your friends are already worm converts and you don't want to sell it, I'm pretty sure there are public parks somewhere within walking distance that could use a few clandestine poop applications. Whatever you do, please don't recycle less garbage because you think you will have too much poop. There is no such thing.

Bartending for Plants:

Turning Worm Poop into Worm Tea

Ahhh, the wondrous world of poop tea! Also called worm tea or compost tea, this is one of the most wonderful parts about worm poop as far as I'm concerned. Poop tea is a great way to be able to put a small amount of worm poop to a much larger purpose. It's like a worm poop extender. But it isn't just for when you have a greater need for worm poop than you have actual poop, you can use the tea to your benefit even if you have an unlimited supply of worm poop at your disposal, like me.

The short story on tea is that, essentially, you are putting the beneficial bacteria and nutrients found in the worm poop into liquid form to make it even more readily available to your plants.

You know when you go to the doctor cause you are really, really, really sick with some bug that is kicking your butt and the only thing you aren't sure of is if you want to crawl into a hole and die or crawl under a rock and die? Well, sometimes your kind doctor *doesn't* tell you it's "just a virus" and that it will pass in another two weeks or two months if you drink lots of water and get some rest. Sometimes your kind doctor will give you an antibiotic to make you feel all better. And *sometimes* your very kind doctor will give you an antibiotic *shot* in addition to the antibiotic pills. And why? So you can get an immediate booster from the shot and the pills are to keep the good times rolling and get your system back in proper working order. Worm tea is a lot like that, except that, unlike antibiotic use, you can use worm tea as often as you want with absolutely no ill effects.

People use worm tea to provide both a boost to their plants and to provide regular nutrition for both in-ground and container grown plants. But the good times don't stop rolling there! The tea is also used as a foliage spray to both cure foliar diseases and fungal issues and to prevent either from taking hold in the first place. It can even be used as an inoculant for your bean, squash and melon seeds before you plant them. It is great to use as a soak for the root ball of both trees and plant seedlings just before you plant them to reduce root shock. Basically worm tea is good for everything except making cake.

The theory on how it works so well, as I understand it, is that the bacteria and microbes in the worm poop are transferred to the water you make the tea with. When the tea is sprayed on the plants, the leaves are covered in the good bacteria and microbes. Depending on what source I'm reading, the good bacteria either displace and overwhelm the bad bacteria and fungi or so cover the leaves that the bad bacteria or fungi don't have a hospitable environment on which to take hold of the leaf. Same difference if you're a bad something-or-other about to infect a plant.

When this same tea mixture is used as a drench to water the plants, the same good bacteria and microbes get rapidly into the soil where the roots can have immediate access to deliver them to the body of the plant, much like how that shot your very nice doctor might give you works in your body. Except in this case, the bacteria and microbes help your plant to make its own hormones and natural defenses to repel the bad things (bad bacteria, fungus, insects) and attract the good things (bees and other beneficial insects) that help them to grow big and strong. It also helps plants recover if they have already been attacked by something that hurt them.

Where I live, we have some wicked hail storms. Usually it is just little hail. But sometimes you get the kind of hail that news crews come out to film, it's so big and nasty. As luck and life would have it, when the bad hail storms do come, it is almost always just after I get my garden all planted. One year in late June, well into growing season here, one of the big and nasty hail storms came through my farm when I was at work. I left that day with a garden full of lush, green plants that were thriving. I came home to a bunch of green sticks. The hail had stripped all the leaves from my plants. Every. Single. One. I won't even begin to describe the trees.

After a wee cry, I whipped up a BUNCH of worm tea. Lots of tea and more than a few prayers went to every green stick. I was still pretty new to the uses of worm tea and not quite sure it would work in the face of such massive destruction. So I bit the bullet and went to the greenhouse and bought some new plants so I could at least have a tomato or two before summer was gone. What a waste.

A waste of money on the new plants, that is. It literally only took a day before the green sticks started getting tiny new leaves. Within two weeks, the plants were back at the "thriving" stage and before summer was out, all of the surviving plants were producing an abundance of fruit. I won't lie to you, neither tea nor poop will bring the truly dead back to life and I lost about four or five plants completely. But the rest? They were gorgeous and productive and none the worse for the trauma they suffered. Tea ROCKS.

Making Tea the Easy-Peasy Way

As with so many other things about raising worms, there are several different ways to make this wonderful and magical tea. Some ways are easy and fast, some ways take a bit of work and finesse and some ways even cost money. You can already guess which way I like best, but naturally we will cover what I know about the other methods so you can make your own informed decision.

The easiest way to make tea is to just take two to five (or so—it's a very scientific process) handfuls of poop, put them in a three to five gallon bucket, add cool or tepid water and mix it up. I have found that it is best to wait at least a few hours or up to two whole days before you use it. Waiting gives the poop time to give up its good stuff to the water. If you wait though, you do have to remember to stir the mixture up every few hours. As you will recall, air is an important component to the whole process of making the poop so that the anaerobic bacteria don't take over and sour the whole process. Same concept with the tea, so make sure you give it a good stir every now and again.

When the tea is "done", your water will look brownish, much like the tea you drink looks. Some of the poop will be dissolved, but there will still be pieces of poop in the tea. This is totally fine, just water your plants with the pieces in the water, it's all good.

Sometimes you may start to make some tea and then life gets in the way the way it likes to do and you may not get the chance to use the tea within a couple of days of making it. Heck, life may keep you from even stirring it regularly. If that happens don't fret too much. If you wait too long or don't stir the mixture to add oxygen, the good bacteria and microbes will start to die off and, if you wait long enough, the bad bacteria will start to grow in the anaerobic environment and start to cause a stinky mess. If you catch it soon enough, just add some fresh water and stir it up real well and soon enough the situation will correct itself and you can use the tea. If too much time has passed and the whole mixture makes you want to puke because it smells so badly, then it is better to start all over from scratch. Frankly, I don't know for sure that it would actively hurt your plants to use it, but I doubt it would actually be helpful. If you have a compost pile, dump the bad stuff into the pile and everything else in the pile will turn the bad into the good. No compost pile? A corner of the lawn or a rocky area will get rid of the bad tea and give you an empty bucket to start all over with.

If you want to make some tea that is not, um, let's say "chunky", you can make the tea with a poop tea bag instead. All you need to do is find some cheesecloth or some super cheap, low thread count fabric and some string to make a tea bag with. I've found that using more tightly woven fabric or even landscape cloth does work, but doesn't work as well. Take a big square (or circle) of the fabric (or even an old, cheapo pillowcase from the thrift store), put however much poop you want to use in the center (or inside the pillow case), then gather up the corners and tie them together with the string. Tada! A tea bag. Put the tea bag in the water and wait. It does take longer, usually at least a whole day, for the tea bag to turn the water into worm tea. You will still want to stir occasionally and I like to give the bag a squeeze or three every now and again to make things work a bit faster. The benefit of using a poop tea bag is that the finished tea can easily be put in a sprayer so you can then apply it to the leaves of your plants without clogging up the sprayer with chunks of worm poop. Of course, you can make it with the bucket method and just strain it well before you put it in your sprayer, too, but why go to all that extra work when a bag works just fine?

Empty the contents of the tea bag onto your garden plants or lawn when you are done. Much like the beverage type of tea bag, trying to use it twice will only give you a very weak tea. But the contents of the bag will still have some good stuff left in it, especially any cocoons that are in the poop, so please don't waste it.

The Harder-Than-it-Needs-to-be Way

As you just read, these passive methods of making tea are almost as easy as falling asleep. There are a couple of other ways to make tea though that you may be interested in if "cheap, easy and fast" just doesn't work for you.

One of the non-cheap-easy-fast ways to make tea is to make what is commonly called aerated compost tea or aerated vermicompost tea, often shortened to either ACT or AVCT. Just like the name says, this is worm tea that you add air to in order to increase the populations of the good bacteria and microbes. Yes, yes, you *did* add air to the passively made tea when you stirred it, but AVCT is made with a LOT more air than that.

AVCT is made with some type of air delivery system, usually like the kind of tubing and small air pump you use if you have goldfish and need to add air to the tank; a tank aerator or what some people call a bubbler. Pumping air into the tea like this provides more oxygen than if you simply stir the tea. More oxygen does help increase the good bacteria and microbes, but not all by itself. In order for AVCT to "work", you not only need to add air but you need to add food for the bacteria. Most recipes call for adding either sugar or molasses in order to give the bacteria and microbes something to eat.

When done properly, the populations of good bacteria and microbes *explode*. The tea goes from millions of good things in the passive system to billions of good things in the AVCT system. Awesome, right? Absolutely. But only if you do it right. And, of course, that is part of the problem. Some people make AVCT and do it right every time and are crazy happy with the whole set up. Others, however, don't do it right every time and end up with a foaming and frothing bucket or tub full of disgusting brown gook that spills over the sides and gets on everything and smells like the devil's own rear end. When I say "others" have this problem, what I mean is "most people". The problem is that there are many, many different recipes and your vermicompost is not necessarily going to be exactly like everyone else's because of what you put into it and how you maintain it. So, say, a cupful of molasses that someone else might use to feed their AVCT made with *their* worm poop (and its attendant bacterial population) might not react the same way as the same recipe made with *your* worm poop. Make sense? Plus, with this system you have to go out and *buy stuff*. That's no fun. Use that money to grow your retirement account instead, okay?

Why, then, do people do this if it is such a waste of time and money, you ask? The theory is that if some good bacteria and microbes are good, a lot more is going to be that much better. "Take two, they're small" taken to the nth degree. While I've read studies that show that worm tea applied as a foliage spray is hugely beneficial, I've not read any studies comparing the results of passively made tea to AVCT to achieve the beneficial effects. As a spray, it does make sense.

As a tea to water your plants with, I'm not sure it does. As a foliar spray, the beneficial bacteria and microbes are sprayed onto the leaves, where they do their good work by basically taking over the place. In the soil though, the roots can only take up so much at a time. How much? I don't know. But I do know that plants grow better when the worm poop is *combined* with the soil, not when the worm poop (and the good microscopic things in the poop) *is* the soil. No, I don't have any studies to back this up. So far I haven't found any. All I've got is logic and experience watching how well regular tea makes my plants grow. Certainly the decision is up to you and the amount of time you want to invest. AVCT won't hurt your plants. If they can't use up all the goodies in the tea, the "goodies" will just die off until they reach a sustainable population level for the conditions in your soil. No harm done, except to your pocketbook and the extra time you wasted making AVCT instead of regular tea in a bucket.

Still not convinced? Well then you can *really* spend some money to make AVCT and buy a tea maker. Yep, you read that right, an actual brewer to make compost tea. Really, I don't make these things up. Depending on the system, expect to drop either a couple or up to several hundred dollars if you want to buy a tea maker. The concept is completely the same as doing it yourself, but this is a fancy system to do all the air adjustments and whatnot for you. Nope, I've never used one and can't imagine I ever will. You may know someone who has one and they totally think it is the cat's meow. Personally, I'm going to use that money for a greater purpose and just make my tea in a bucket. I've got a farm full of gorgeous green to show for my frugality so you'll hear no complaints from me!

Well, you maybe never thought I'd say this but: that's it! I've tried to cover everything without overwhelming you with every little thing and I hope it's been fun for you, too. But knowing you I bet there is still a question or two you would like to ask or an experience you would like to share. Please feel free to do so (within reason of course!) by emailing me at:

WiRFarm@yahoo.com (no attachments, please) or find me on facebook at "The Best Place for Garbage". I'll include all of the good questions (and answers) and stories in a follow-up book and that way you will finally know the poop, the whole poop, and nothing but. I also love to get worm comics and funny worm jokes...Hey, you never know, you might have one I've never heard before!

Regardless of what type of system you use to grow your worms in or how you use the poop, the important thing is that you get started. With even conservative estimates placing the proportion of organic recyclables in our household waste stream at about 30% (Wow! And around **60%** of *all* waste—double wow!), every little bit you can convert *really does count*. Please don't just "throw it away" unless you first give some serious thought as to just what "away" really means.

Thank you so much for recycling your organic waste and don't forget to spread the word...I mean *worm!*

EXPERIMENTS IN POOP

You may recall that I mentioned once or umpteen times that there is a lot of room for some more definitive research in some areas associated with worm composting and nothing but room for *any* kind of research in other areas. I also mentioned that there are businesses and municipalities and ordinary people just like you and me who have been vermicomposting for ages. Too few of these folks have been keeping track of exactly what they are doing and even fewer have been writing about it.

One way that "novel" ideas (like composting with worms, just as a 'for instance') get to be more accepted is by having proven research to back up anecdotal evidence. Anecdotal evidence is the kind of information you gather from doing stuff but without doing it according to a bunch of set rules and keeping accurate records of what you are doing. Most of what is known about raising composting worms is, unfortunately, only anecdotal. A lot of anecdotal evidence that all has the same result is helpful but will rarely meet the level of solid, by-the-rules, writing-everything-down research.

Solid research not only helps for ordinary folks like you and me, but goes a very long way in convincing larger bodies of people, like governments and agricultural interests and other large institutions. While it is unlikely that this book will ever get to be an Oprah Winfrey "Book of the Month" (but wouldn't that be *cool*?), the more you and I "spread the worm" *and* back that up with solid research, the more garbage we can not only keep out of the landfills, but the more nutrients we can put back into the soil where we can get them back into our bodies through our food.

In the not too distant past, no one blinked when your favorite restaurant had a huge dumpster in the back where every scrap of waste was dumped. Now, it is totally normal for that same restaurant to have a smaller dumpster and a recycle bin for bottles and cans and cardboard. A lot of restaurants also now have a recycling container for old fry oil. Can you imagine how great the day will be when that same restaurant has an even smaller (teensy weensy would be good) dumpster because they also have a container for food waste? Sigh. There I go dreaming again...But it is a totally worthwhile dream and believe me when I say I am definitely not the only person who would benefit from that dream coming true.

So, how do we get closer to making that dream a reality? There are several avenues. First, of course, is your inability to keep quiet about the whole worm poop thing. The next thing would be increasing the amount of research being done not only in the area of earthworm composting, but also in the areas of regular composting and soil science and plant growth and the nutritional content of the resulting food. All are inextricably related and research in one area inevitably benefits the others.

How do you increase the amount of research when you are just an ordinary Joe? That answer is almost as easy as worm composting itself and, just like everything else, there are several means to this end. One easy way is to familiarize yourself with how to conduct simple experiments. There are certain minimum requirements and any junior high school science book will help in that area. Hey, you were going to help your kids with their homework anyway, right? Even conducting simple experiments with your own worms and keeping accurate records helps. But only if the experiment is conducted right *and* only if you let other people know about it. I can help in that second area. I'd like to put everything *you* learn experimenting with your worms in a central place, like a sequel to this book. No, it doesn't pay well so don't get your hopes up on that front. But it definitely will help the whole entire planet in the long run and how many things do you get to do in the privacy of your own home that can meet that lofty goal? So experiment in whatever area piques your interest and do keep records and write it all down and send it to me when you are ready. You get total credit and the undying thanks of a grateful nation...Okay, okay, you will at least get full credit.

Another thing you can do, if you are enrolled as a student in any learning institution, is to conduct experiments as part of your class work or thesis research. Generally you will have an advantage in this area because you will have access to mentors who can help you design bigger and more specific experiments than you will most likely be able to do all on your own. You will also, if you do it right and ask nicely, get the opportunity to publish the results of that research in some of the hoity-toity soil science journals. But that doesn't mean you also can't send it to me! Please do.

Lastly, there are two things that those of you who don't have a burning desire to do detailed research can still do to help. The main thing you can do is help to both encourage and *sponsor* research in any of these areas. The unfortunate reality is that the vast majority of the research that gets done these days is done because someone who is interested in the outcome puts their money where their mouth is and either pays for or helps to pay for the research. You can do this in large ways and small, you don't have to be a billionaire philanthropist to participate. Contact the nearest university with a soil science program and ask them how you can help. Trust me, they'll find a way. From actually funding research to contributing to scholarships or even 'just' helping to put together a composting program for the school cafeteria, you can be a part of what will make the crucial difference in saving valuable waste from becoming just waste.

Another thing you can do that might seem silly and inconsequential on the surface but is still totally helpful even if you don't ever get to witness the final outcome: you can encourage young students to explore the amazing world of soil science. Teach your kids or grandkids about gardening and farming. No kids? I bet your neighbors have some you can borrow. Hate kids? You can still talk to their parents about the wonderful world of dirt! Make it fun, make it exciting! They'll listen if you present it right. Don't say that is a ridiculous idea, after all, you read this whole book and liked it! Admit it, you did. Are you, by chance, a famous movie or sports star? Maybe you could do a public service announcement convincing people that studying soil science is waaaaaaaaay sexier than studying rocket science! Hey, you know how people are influenced by popular culture, give it a shot before you laugh it off. The point of all of this is that you don't always know the impact of your positive words. You likely didn't know about all this worm composting business until someone else you like told you about it. And now look at you!

Don't worry, I won't leave you hanging, trying to come up with all the good ideas on your own. You can totally come up with good ideas outside of this list, but here are a few areas of research that are sorely lacking in hard data and, at a minimum, will get the ol' gray matter rocking and rolling on ways to find out more.

(1) What happens if you feed worms poisonous plants like poison ivy, poison sumac, castor bean plants and the like? Does the poison hurt the worms? Are the poisonous properties of the plant rendered innocuous to humans or is it still present? If it is still present, would the poison be transferred to the fruit if the resulting worm poop is used? The exact same question applies to feeding worms a steady diet of plants infected with disease or fungus or the like and deceased animals infected with diseases.

(2) What happens if you feed worms sawdust and chips made from treated, stained or painted wood? Is there an acceptable percentage? Are any of these treatments rendered innocuous by the action of the worms? Are toxins present in the worm poop? If so, are they taken up by the plants or fruits of the plant?

(3) More experimentation is needed regarding the effects of seeding worm beds with harmful bacteria like salmonella and the like. Particularly, more research is needed to find out if the mitigating effects of the worm poop are sustained over time or if there is an upper limit the worms can't overcome and also if the worm poop has a mitigating effect on harmful bacteria when it is used as part of the soil mixture as opposed to seeding worm beds directly (meaning: added to soil that has been 'infected'). Also, what if harmful bacteria is seeded into soil that is already high in organic matter and has abundant earthworms and microbial life? Testing of the plants and fruits in all of these areas would also be beneficial.

(4) Does spraying worm tea on the leaves of plants that have been recently treated with unfinished compost and manures have an effect on the population of harmful bacteria in the unfinished matter? Particular attention in this area must be paid to the underside of the leaves as well as the top sides due to backsplash.

(5) Design an inexpensive electric sifter that can be powered by alternative energy like solar energy or maybe even by a bicycle or the like. Something cheap and easy for anyone to make. Making it operable by one person alone would be ideal. Designing a sifter that could easily remove plastic contamination would also be highly beneficial.

(6) Is there a difference in the measurable nutritional content of worm poop fed specific diets? Is the poop from worms fed solely on coffee grounds significantly different from worms fed a varied diet of food scraps? What about worms fed commercially prepared worm chow? Treated or raw sewage? Horse or cow manures? Manures from confined livestock feeding operations (CAFO's)? Brewery waste? Winery waste? Waste from an orange juice processor? Spaghetti sauce

plant? Silage from an organic grain field? What about silage from non-organic farms? Does it really make a difference or does the miniature miracle of worm digestion make the poop the same regardless of what they are fed?

(7) Does using dirty, soapy dishwater have any effect on the worm population or the bacterial population (presuming non-anti-bacterial soap)? Any effect on the poop itself?

(8) Does using aerated vermicompost tea (AVCT) have any more beneficial effects than using passively made tea? Although AVCT has been shown to have higher populations of beneficial bacteria, does it really matter in fighting plant diseases and fungus or is one just as good as the other as far as the plant is concerned? Same question whether the teas are sprayed on or used as a soil drench. The same question would also apply to (4) regarding treating plants that have been sprayed with unfinished compost or manures.

(9) How do the chemical/physical properties of worm poop change over time if stored? Stored with water added or stored dry? Stored in the cold or the heat? Stored for longer than six months? Same questions would apply to pre-made teas that are sometimes sold as plant fertilizer.

See how your mind is working on other things already and you haven't even picked up a notebook to write down your experiment ideas? The internet is actually a great place to search for related research to help you refine your study ideas before you start or even just to do research on the research for the sake of knowledge alone. Do remember a few things, though, when reading about these studies and experiments:

-just because a study wasn't done at a university or some such doesn't mean it isn't valid or worthwhile

-likewise just because a study *was* done at a university or similar institution doesn't mean it *is* valid or worthwhile

-a lot of 'research' has been done or sponsored by companies with a vested interest in the outcome, such as chemical fertilizer companies; be aware of who is sponsoring the research and who is reporting the results

-far, far more research has been done regarding "conventional" versus organic *large-scale* agricultural practices and much of this research has been blindly applied to small scale agricultural and gardening practices even if what happens on a 2000 acre farm is not the same as what happens in your 20 x 10 foot garden

-if you are reading about research results that give you an uneasy feeling in your gut and you aren't sure why, trust your feelings; you   may not yet know what is wrong with the research, but the more education you give yourself in these areas, the better prepared you are to ferret out the crap from the good poop

The most important thing is that you start or help someone else who can. The stakes are huge. Even though I seem to have worm poop on the brain, the real issues here are the health of our physical environment and the health of the food we eat. What trumps that?

INDEX

235

ABOUT THE AUTHOR

Sandra (Sandi) Wiese is a small-scale commercial worm composter with one simple dream: that someday "I have worms" will not be met with a cringe and "Well I hope the doctor can give you something for them!". Among other things, she runs a small organics recycling business on her farm where she also produces livestock food, people food and, of course, worms.
She lives in Bennett, Colorado with her Rabrador letreiver and assorted other misfits.

21886177R00152

Made in the USA
San Bernardino, CA
10 June 2015